Lecture Notes in Computer Science 9987

Commenced Publication in 1973
Founding and Former Series Editors:
Gerhard Goos, Juris Hartmanis, and Jan van Leeuwen

More information about this series at http://www.springer.com/series/7410

Barbara Kordy · Mathias Ekstedt
Dong Seong Kim (Eds.)

Graphical Models for Security

Third International Workshop, GraMSec 2016
Lisbon, Portugal, June 27, 2016
Revised Selected Papers

 Springer

Editors
Barbara Kordy
INSA Rennes, IRISA
Rennes
France

Mathias Ekstedt
KTH Royal Institute of Technology
Stockholm
Sweden

Dong Seong Kim
University of Canterbury
Christchurch
New Zealand

ISSN 0302-9743   ISSN 1611-3349 (electronic)
Lecture Notes in Computer Science
ISBN 978-3-319-46262-2   ISBN 978-3-319-46263-9 (eBook)
DOI 10.1007/978-3-319-46263-9

Library of Congress Control Number: 2016951653

LNCS Sublibrary: SL4 – Security and Cryptology

Printed on acid-free paper

This Springer imprint is published by Springer Nature
The registered company is Springer International Publishing AG
The registered company address is: Gewerbestrasse 11, 6330 Cham, Switzerland

Preface

The present volume contains the proceedings of the Third International Workshop on Graphical Models for Security (GraMSec 2016). The workshop was held in Lisbon, Portugal, on June 27, 2016, in conjunction with the 29th IEEE Computer Security Foundations Symposium (CSF 2016).

Using graphical security models to represent and analyze the security of systems has gained increasing attention over the last two decades. Graphical models are used to capture different security facets and address a range of challenges, including security assessment, automated defensing, secure services composition, security policy validation and verification. GraMSec brings together academic researchers as well as industry and government practitioners designing and employing visual models for security. It creates a platform for the exchange of ideas, discussion, inspiration, collaboration, and dissemination of results in the field of graphical security modeling. It contributes to the development of well-founded graphical security models, efficient algorithms for their analysis, as well as methodologies for their practical usage.

GraMSec 2016 received 23 submissions, which represents a growth of 77 % compared with the first and the second edition of the workshop. The papers are co-authored by experts from 18 countries. Each article was reviewed by at least three reviewers. Based on their quality and contribution to the field, nine papers were accepted for presentation at the workshop and inclusion in the final proceedings. The technical program was complemented by an invited talk by Xinming Ou, entitled "Bottom-Up Approach to Applying Graphical Models in Security Analysis." The corresponding invited paper has been included in these proceedings.

We would like to express our deepest appreciation to all the people who volunteered their time and energy to make this year's workshop happen. In particular, we thank the authors for submitting their manuscripts to the workshop and all the attendees for contributing to the workshop discussions. We are also grateful to the members of the Program Committee and the external reviewers for their work in evaluating and discussing the submissions, and their commitment to meeting the strict deadlines. A very special recognition is dedicated to Pedro Adão — the General Chair of CSF 2016 — for his invaluable support in organizing GraMSec 2016.

Our thanks also go to the European Commission's Seventh Framework Programme (EU FP7 grant no. 318003 TREsPASS), the University of Luxembourg, the Fonds National de la Recherche Luxembourg (FNR-CORE grant ADT2P), and INSA Rennes for their partial sponsorship of the workshop, as well as KTH Royal Institute of Technology and the IRISA institute for their in kind contribution to GraMSec 2016.

Finally, we would like to acknowledge Springer for accepting to publish these proceedings as an LNCS volume as well as the EasyChair team for providing a very practical tool supporting the workshop's management and the preparation of these proceedings.

August 2016   Barbara Kordy
                                                        Mathias Ekstedt
                                                        Dong Seong Kim

Organization

Program Committee

Mathieu Acher	University Rennes 1/Inria, France
Massimiliano Albanese	George Mason University, USA
Ludovic Apvrille	Telecom ParisTech, France
Thomas Bauereiss	DFKI, Germany
Giampaolo Bella	Università di Catania, Italy
Stefano Bistarelli	Università di Perugia, Italy
Marc Bouissou	EDF, France
Binbin Chen	Advanced Digital Sciences Center, Singapore
Jason Crampton	Royal Holloway, University of London, UK
Frédéric Cuppens	Télécom Bretagne, France
Nora Cuppens-Boulahia	Télécom Bretagne, France
Hervé Debar	Télécom SudParis, France
Giovanna Dondossola	RSE, Italy
Mathias Ekstedt	KTH - Royal Institute of Technology, Sweden
Ulrik Franke	Swedish Institute of Computer Science - SICS, Sweden
Frank Fransen	TNO, The Netherlands
Olga Gadyatskaya	SnT, University of Luxembourg, Luxembourg
Paolo Giorgini	University of Trento, Italy
Erlend Andreas Gjære	SINTEF, Norway
Dieter Gollmann	Hamburg University of Technology, Germany
René Rydhof Hansen	Aalborg University, Denmark
Olivier Heen	Inria/Thomson, France
Hannes Holm	Swedish Defence Research Agency, Sweden
Siv Hilde Houmb	Secure-NOK AS, Norway
Sushil Jajodia	George Mason University, USA
Lanet Jean-Louis	Inria-RBA, France
Ravi Jhawar	University of Luxembourg, Luxembourg
Henk Jonkers	BiZZdesign, The Netherlands
Florian Kammueller	Middlesex University London, UK
Nima Khakzad	TU Delft, The Netherlands
Dong Seong Kim	University of Canterbury, New Zealand
Barbara Kordy	INSA Rennes, IRISA, France
Pascal Lafourcade	VERIMAG, University of Grenoble, France
Jean Leneutre	Ecole Nationale Supérieure des Télécommunications (ENST), France
David Lubicz	DGA-MI, University of Rennes 1, France
Sjouke Mauw	University of Luxembourg, Luxembourg

Per Håkon Meland	SINTEF ICT, Norway
Jogesh Muppala	HKUST, Hong Kong, SAR China
Simin Nadjm-Tehrani	Linköping University, Sweden
Steven Noel	MITRE, USA
Andreas L. Opdahl	University of Bergen, Norway
Xinming Ou	University of South Florida, USA
Stéphane Paul	Thales Research and Technology, France
Ludovic Piètre-Cambacédès	EDF, France
Sophie Pinchinat	IRISA, University of Rennes 1, France
Vincenzo Piuri	University of Milan, Italy
Marc Pouly	Lucerne University of Applied Sciences and Arts, Switzerland
Nicolas Prigent	Supélec, France
Cristian Prisacariu	University of Oslo, Norway
Christian W. Probst	Technical University of Denmark, Denmark
David Pym	UCL, UK
Saša Radomirovic	ETH Zurich, Switzerland
Indrajit Ray	Colorado State University, USA
Frédéric Remi	AMOSSYS, France
Arend Rensink	University of Twente, The Netherlands
Yves Roudier	EURECOM, France
Pierangela Samarati	Università degli Studi di Milano, Italy
Guttorm Sindre	NTNU, Norway
Ketil Stølen	SINTEF, Norway
Mariëlle I.A. Stoelinga	University of Twente, The Netherlands
Axel Tanner	IBM Research - Zurich, Switzerland
Kishor Trivedi	Duke University, USA
Alexandre Vernotte	KTH — Royal Institute of Technology, Sweden
Luca Viganò	King's College London, UK
Lingyu Wang	Concordia University, Canada
Jan Willemson	Cybernetica, Estonia

Additional Reviewers

Bertrand, Nathalie	Gharib, Mohamad
Chang, Xiaolin	Li, Letitia
Erdogan, Gencer	Venkatesan, Sridhar
Genet, Thomas	

Contents

A Bottom-Up Approach to Applying Graphical Models in Security Analysis

Xinming Ou[✉]

University of South Florida, Tampa, USA
xou@usf.edu

Abstract. Graphical models have emerged as a widely adopted approach to conducting security analysis for computer and network systems. The power of graphical models lies in two aspects: the graph structure can be used to capture correlations among security events, and the quantitative reasoning over the graph structure can render useful triaging decisions when dealing with the inherent uncertainty in security events. In this work we leverage these powers afforded by graphical model in security analysis. Given that the analyst is the intended user of the model, the most difficult task for research in this area is to understand the real world constraints under which security analysts must operate with. Those constraints dictate what parameters are realistically obtainable to use in the designed graphical models, and what type of reasoning results can be useful to analysts. We present how we use this bottom-up approach to design customized graphical models for enterprise network intrusion analysis. In this work, we had to design specific graph generation algorithms based on the concrete security problems at hands, and customized reasoning algorithms to use the graphical model to yield useful tools for analysts.

1 Introduction

Intrusion analysis is the process of examining real-time events such as IDS alerts and audit logs to identify and confirm successful attacks and attack attempts into computer systems. The IDS sensors that we have to rely on for this purpose often suffer from a large false positive rate. For example, we run the well-known open-source IDS system Snort on our departmental network containing just a couple of hundreds machines and Snort produces hundreds of thousands of alerts every day, most of which happen to be false alarms. The reason for this are well-known: to prevent false negatives, *i.e.* detection misses from overly specific attack signatures, the signatures that are loaded in the IDS are often as general as possible, so that an activity with even a remote possibility of indicating an attack will trigger an alert. It then becomes the responsibility of a human analyst monitoring the IDS system to distinguish the true alarms from the enormous number of false ones. How to deal with the overwhelming prevalence of false positives is the primary challenge in making IDS sensors useful, as pointed out by Axelsson [3] more than 10 years ago.

This article was based on a previously published work [38].

© Springer International Publishing AG 2016
B. Kordy et al. (Eds.): GraMSec 2016, LNCS 9987, pp. 1–24, 2016.
DOI: 10.1007/978-3-319-46263-9_1

Due to the lack of effective techniques to handle the false-positive problem, it is common among practitioners to altogether disable IDS signatures that tend to trigger large number of false positives. At one site we visited, the security analysts did not use the standard Snort rule sets at all, but rather resorted to secret, i.e. unpublished, attack signatures that are highly specific to their experience and environment and with known (small) false negative rates. We were told by the security analysts that secret signatures can only help capture some "low-hanging fruit", and that many attacks are likely missed due to the disabled signatures. Turning off IDS signatures is like turning a blind eye to attack possibilities, which we believe is a drastic consequence of the lack of effective solution techniques to *prioritize* investigations of alerts from IDS and audit logs. But, lacking any other significant distinguishing feature between the alerts, practitioners see no alternative.

1.1 Quantifying Uncertainty

Current IDS systems do not distinguish nor help distinguish the alarms that are highly likely to be true from those that have only a small chance of being true. By treating each suspected or imputed attack as has been suggested in earlier literature (see, e.g. [5] and references therein), merely as a hypothesis whose validity needs to be established, an effective approach to dealing with false positives can be formulated. The task then is to quantify the uncertainty in the hypotheses ascribed to IDS alerts by correlating multiple-point observations that are relevant to each alert. Given a list of intrusion hypotheses sorted by confidence and annotated by the evidential support for each hypothesis, it would be much easier for a human analyst to decide which hypotheses deserve further investigation. Since most network intrusions involve multiple actions, if we can relate observations from multiple events, a true successful attack will likely have multiple pieces of corroborating evidence, thus increasing the certainty of the attack hypothesis. Correspondingly, a false positive in one sensor is likely to have less corroborating evidence, thus the particular attack hypothesis will have a low score and be ignored. The key question then is *how to calculate a difficult formulation based on both the reasoning structure in which it is derived and the quality of the evidence that supports it.*

There have been past attempts [34, 36] at achieving this. Bayesian analysis [14] has been the standard and there have been some approaches using alternative theories such as Dempster-Shafer theory [23]. However, a number of **fundamental issues** in applying these mathematical theories to intrusion analysis remain to be addressed. For Bayesian analysis, it seems difficult to establish adequate prior probabilities such as the probability of a specific attack occurring in the environment or determine the conditional probabilities between system events in a robust manner. For Dempster-Shafer theory, it is not clear how to model sensor quality, where to obtain such parameters, and how to handle non-independent sources of evidence.

1.2 Our Contributions

Dempster-Shafer theory has unique advantages in handling uncertainty in intrusion analysis, namely, the ability to deal with the lack of prior probabilities for all (singleton) events and the ability to combine beliefs from multiple sources of evidence [6,7,34]. In this paper we present an extended Dempster-Shafer model that addresses the fundamental issues in applying DS in intrusion analysis, as mentioned in Sect. 1.1. We have implemented our method on top of our IDS alert correlation tool SnIPS [15,21], so that one can calculate a numeric confidence score for each derived hypothesis and prioritize the results based on the scores. Our contributions are:

Using "unknown" to capture sensor quality (see Sect. 3.2). Dempster-Shafer theory allows specifying a weight on "unknown" (or "to be determined") rather than specifying precise probabilities for every possible event in the space. We use this ability to represent lack of knowledge to capture the intuitive notion of IDS sensor quality (which usually turns out to be imprecisely described), without suffering the non-intuitive effects of aggregation or forced classification that have been observed by researchers [34].

Accounting for lack of independence among alerts (see Sect. 3.3). A long-standing assumption in DS theory is that multiple pieces of evidence are independent, which is a property that is hard to confirm in practice. This is especially a problem in IDS alerts since many alerts are triggered by the same or similar signatures. In combining these alerts to derive the overall belief on the attack status, it is important that such non-independence be appropriately accounted for so that the result is not skewed by over-counting. To the best of our knowledge, our method is the first in applying sound non-independent DS belief combination in IDS alerts.

Efficient calculation (see Appendix B). A direct application of DS formulas can result in exponential (in the number of hypotheses – in our case, IP addresses) blow-up of belief combinations. We adopt a *"translate-then-combine"* approach so that beliefs are propagated in a correlation graph and only combined at join points in the graph. This produces an efficient algorithm with worst-case running time quadratic in the number of IP addresses in the input alerts.

Linking to practical IDS tools (see Sect. 4). We have implemented our approach for the open-source IDS system Snort, and applied it on a production network and a number of data sets. We rigorously evaluate our method by separating the tuning phase from the testing phase, so that we do not fit our tool's parameters to work well with just one particular data set. Our evaluation suggests that the scores computed from our algorithm provide a useful ranking for the correlated alerts based on the correlations' trustworthiness. We have validated the results both anecdotally as well as with data set ground truths whenever available.

Robustness of solution (see Sect. 4.4). We emphasize that our final goal is to sort alerts by confidence, hence we are interested in the relative order of hypotheses by confidence, not in establishing absolute certainties about attacks. Our application of DS requires the assignment of numeric values (constants) to certainty levels {unlikely, possible, likely, probable} but there is no help in the theory itself as to the manner of assignment. In a standard application of DS, the numeric scores may affect the final conclusions. However, since we are interested only in *relative* belief strengths assigned to the hypotheses, our approach is robust to small changes in these constants. Given any two related hypotheses, the absolute belief values are irrelevant as long as the relative strengths of belief remain unchanged when we slightly vary the numeric constants. Our experimental analysis shows that this is indeed true; the classifier's operating characteristic does not change when the constants' values are varied within a small range.

2   Background on Dempster-Shafer Theory

A common example to illustrate the difference between probability theory and Dempster-Shafer theory is that if we toss a coin with an unknown bias, probability theory will still assign 50 % for Heads and 50 % for Tails by the principle of indifference ([11], p. 167), which states that all states of unknown probability should be assigned equal probability. Dempster-Shafer theory, on the other hand, handles this by assigning 0 % belief to {*Head*} and {*Tail*} and assigning 100 % belief to the *set* {*Head, Tail*}, meaning "either Head or Tail". By allowing us to assign 100 % belief to {*Head, Tail*} if warranted, DS does not force us to pick a probability when there is no basis to assign it. More generally, the DS approach allows for three kinds of answers: *Yes, No, or Don't know*. The last option of allowing ignorance makes a big difference in evidential reasoning. See [13], Chap. 2 for a discussion of the relative merits of DS belief theory. In DS theory, a set of disjoint hypotheses of interest, *e.g.*, {*attack, no-attack*}, is called a *frame of discernment*. The *basic probability assignment*, (*bpa*) function, also called the *mass distribution function* (m_θ), distributes the belief over the *power set* of the frame of discernment and is defined as:

$$m_\theta : 2^\theta \to [0, 1] \tag{1}$$

$$m_\theta(\{\}) = 0 \quad \text{and} \quad \sum_{x \subseteq \theta} m_\theta(x) = 1 \tag{2}$$

Definition 1. *Let θ be a frame of discernment and m_θ a bpa function. The belief function is defined as*

$$Bel(x) = \sum_{y \subseteq x} m_\theta(y), for\ x \subseteq \theta \tag{3}$$

2.1   Dempster's Rule of Combination

The goal of combination is to fuse the evidences for a hypothesis from multiple *independent* sources and calculate an overall belief for the hypothesis [22].

Figure 1 illustrates this idea, where $alert_2$, $alert_3$ are two alerts triggered by independent IDS sensors. Independence means that knowing whether one sensor is trustworthy or not will not influence the likelihood for the other being trustworthy or not. A common assumption is that if two sensors are independent if they operate on completely unrelated features to determine attack possibilities. Both alerts could indicate that machine ip_2 is doing malicious probing of ip_3. The question is how we combine the beliefs from the two evidence sources. In general we have the following rule for fusing known as the Dempster's rule of combination.

$$m_{1,2}(h) = \frac{1}{1-K} \cdot \sum_{h_1 \cap h_2 = h} m_1(h_1) \cdot m_2(h_2) \tag{4}$$

$$K = \sum_{h_1 \cap h_2 = \{\}} m_1(h_1) \cdot m_2(h_2) \tag{5}$$

Where h_i's and h's are subsets of H, hypotheses in the frame of discernment. K is a normalization factor that is a measure of the *conflict* between the two sources of evidence, which is equivalent to the measure of the cases of empty intersection between the h_i's. The combined mass function must be normalized by $1-K$ when conflict exists [22,23].

The multiplication in formula 4 is only valid when the two evidence sources are independent [22]. This is often not the case in practice and especially so in IDS alerts since many alerts are generated by the same or related signatures. In the next section we introduce our extension of the DS model to account for non-independent evidence sources, so that the DS model can be correctly applied in intrusion analysis.

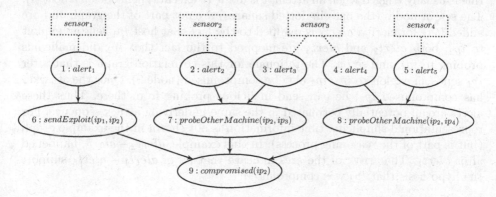

Fig. 1. Automatically generated correlation graph segment from SnIPS

3   Applying Dempster Shafer on Intrusion Analysis

3.1   The SnIPS Framework

We have built our DS-based hypothesis prioritization model on top of the SnIPS [15,21] intrusion analysis system from our past work. SnIPS can work with the open-source Snort IDS system. It maps a triggered IDS alert to a hypothesis such as "machine compromised." It also maps the trustworthiness of the hypothesis to a discrete tag such as "possible" and "likely". Our past work [21] showed that building the mappings does not require much additional work, since the information already exists in an ad-hoc manner in the Snort rule repository. We have developed a heuristic algorithm to automatically infer the mappings by analyzing the Snort rules' documentation. After the alerts are mapped to hypotheses, the hypotheses are reasoned about efficiently based on a succinct internal reasoning model, and an alert correlation graph is built that shows the possible links among the hypotheses and alerts [29].

Example of SnIPS Output. Figure 1 shows a sample segment of alert correlation graph automatically generated by SnIPS. "*compromised*", "*sendExploit*", and "*probeOtherMachine*" are predicates used to describe various attack hypotheses. The arrows' direction in the graph is aligned with inference. Five groups of alerts $alert_1 - alert_5$, are triggered by four different sensors. The notion of a sensor in our model is a bit different than other previous works. In our model we are not using the notion of physical snort sensor (i.e. Network card), or IDS in general. Instead, we are using each snort signature as a virtual sensor supporting the correlation graph. This is under the assumption that snort alerts will be triggered independently. For example in Fig. 1 $sensor_1$ could be snort rule *1:1390*. This rule is usually trigged when an attempt is made to execute shellcode on a host [1]. The sensor nodes (the ones in dotted squares) are not part of the graph and are added here for clarity. $alert_1$ is mapped to the fact that host ip_1 sent an exploit to ip_2; both $alert_2$ and $alert_3$ are mapped to the fact that ip_2 did malicious probing to ip_3, and so on. The rationale for this correlation graph is that after ip_1 sends an exploit to ip_2, ip_2 may be compromised (node 9). Once the attacker has compromised ip_2, he can send malicious probing from there. Thus these alerts are all potentially correlated in the same underlying attack sequence. For representational simplicity, time information is not shown in the example graph (but is part of the reasoning process). In this example, $alert_2 - alert_5$ happened after $alert_1$. The arrow of the arcs indicate that all of $alert_1 - alert_5$ support the hypothesis that ip_2 was compromised.

3.2   Metrics for Sensor Quality

False positive and negative rates have been the standard metrics for characterizing an IDS sensor's quality. In this work we do *not* subscribe to such probabilistic metrics. Rather, we will use the "unknown feature" provided by DS theory to capture the case when we do not trust a sensor. The nature of *unknown* matches naturally with how humans interpret IDS alerts. When an alert is fired, we will have some degree (say 10 %) of belief that an attack is going on. If we were to use a probability interpretation, we would have to say that we have 90 % belief that the attack is *not* going on. One may find it counter-intuitive to positively assert that an attack is not going on based on seeing an alert. In probability theory, this is addressed by comparing the probability of an attack before and after seeing an alert. However, this would require the specification of the prior probability of attacks, which is hard if not impossible to obtain. By using DS, we can assign 0.1 belief to "attack"($\{true\}$), 0 belief to "no-attack" ($\{false\}$), and 0.9 belief to "Don't know" ($\{true, false\}$). This is a more intuitive quantitative interpretation of what an IDS alert provides: it gives some (small) belief that there is an attack but it does *not* give us any belief for "no-attack." Just because the sensor is not trustworthy, does not mean an attack is not going on. There may still be attack that is completely outside the scope of the sensor's detection. Assigning the remaining weight to the "unknown" state indicates that we acknowledge the open-ended nature of attacks, which captures the reality of cyber security.

Using this method, we only need a single metric δ to characterize a sensor's quality. δ corresponds to the bpa of $\{true\}$ for the corresponding hypothesis when the sensor fires. Then $1 - \delta$ will be assigned to $\{true, false\}$ (denoted θ thereafter). In the example of Fig. 1, if we have $\delta = 0.1$ as $sensor_1$'s trustworthiness, $alert_1$ will translate to 0.1 mass distribution for $sendExploit(ip_1, ip_2)$ being *true*. 0.9 weight will be distributed to $sendExploit(ip_1, ip_2)$ being θ ("unknown"). We view δ as a metric solely dependent on the sensor's trustworthiness. We also assume for simplicity that shared IDS sensors only give us positive correlation, i.e. the triggering of one alert cannot cause us to decrease our belief in another correlated alert but only to increase it or stay the same. IDS signatures often come with ad-hoc natural-language descriptions that indicate the quality of the signature in terms of how likely the triggered alerts will be false positives, using qualitative terms such as "possible" and "likely." SnIPS extracts such terms from the Snort rule documentation and assigns a corresponding "certainty tag" for alerts generated by the rule [21]. In practice such tags can be provided easily by the rule writer if they are standardized, since they are already used in an informal way. We use the SnIPS certainty tags to map to the quantitative quality metrics for alerts generated by the various Snort rules (sensors), in the scheme shown in Table 1.

Table 1. Mapping discrete certainty tags to quantitative sensor quality metrics

Measures		Metrics	Measures		Metrics
Unlikely	→	0.01	Possible	→	0.33
Likely	→	0.66	Probable	→	0.99

The intuition is that humans typically cannot distinguish small differences in numerical parameters, thus a few discrete levels are sufficient to express the various beliefs one can ascribe to an alert. Through our analysis of the Snort rules' documentation, we found that four levels are sufficient to differentiate the various belief levels reflected by the rule writers about an alert's trustworthiness [21]. There is a low belief 0.01 and a high belief 0.99. The other two levels are evenly divided in the middle space. Another consequence of this model of sensor quality is that there will be no conflict among alerts. When we do not trust an alert, we just say "Don't know" whether the hypothesis is true, rather than assert that the hypothesis is false. This will not contradict the fact that we may trust another alert which derives the same hypothesis to be true. Thus we always have $K = 0$ in the combination formula (4).

3.3   Extending DS Combination Rule for Non-independent Evidence

Correlated alerts could provide falsely elevated belief that an attack is going on, since multiple pieces of evidence point to the same conclusion. A key question is whether these multiple pieces come from independent sources. Through our research we discovered that we cannot ignore or avoid the overlapping nature of evidence. Often times we see multiple alerts in correlation supporting a hypothesis, but these alerts are triggered by the same or similar IDS signatures leading to an unjustifiably high level of confidence if we apply the standard Dempster rule of combination. In reality these multiple alerts should not significantly increase our belief in the hypothesis.

There may also be "partial non-independence" between two sources of evidence. In Fig. 1, the main hypothesis is node 9: "whether machine ip_2 is compromised." This hypothesis is supported by the alert node 1–5. Node 1, an alert triggered by $sensor_1$, has evidence supporting node 6. Node 2 and 3 have evidence supporting node 7, so we combine the belief in 2 and 3 into node 7. Similarly, the belief in 4 and 5 will be combined into 8. Then we need to combine the belief in 6, 7, and 8 to answer the final question in node 9. Now we cannot ignore the fact that these nodes have overlapping evidence. Specifically, both node 7 and 8 partially rely upon alerts triggered by $sensor_3$. As a result, node 7 and 8 are not completely independent and we cannot simply apply the Dempster rule of combination (Sect. 2.1).

There are a number of approaches in the DS literature to account for such dependence [10, 24–26]. We adopt an idea proposed originally by Shafer [26] which interprets combined bpa's as joint probabilities. Based on this, we develop a set of customized combination formulas to correctly account for the dependence in evidence when combining beliefs in the alert correlation graph.

The Customized Combination Formula. The reason Dempster's rule of combination has to assume evidence sources are independent is that joint mass function is calculated through multiplication (formula 4). For non-independent evidence, multiplication of bpa's from two sources is no longer valid [26]. Instead of $m_1(h_1) \cdot m_2(h_2)$, we use $\psi[h_1, h_2]$ to denote the joint bpa of the two sources. We obtain the following new formula for combining possibly non-independent evidence.

$$m_{1,2}(h) = \sum_{h_1 \cap h_2 = h} \psi[h_1, h_2] \qquad (6)$$

One implication that arises from the application domain, namely intrusion analysis, is that the only h_i's of interest are $\{true\}$ (referred to as t hereafter) and $\{true, false\}$ (referred to as θ hereafter). The following equations specify $\psi[h_1, h_2]$,

$$\psi[t, t] = r_1 \cdot m_1(t) + (1 - r_1) \cdot m_1(t) \cdot m_2(t) \qquad (7)$$

$$\psi[t, \theta] = (1 - r_1) \cdot m_1(t) \cdot m_2(\theta) \qquad (8)$$

$$\psi[\theta, t] = (1 - r_2) \cdot m_1(\theta) \cdot m_2(t) \qquad (9)$$

$$\psi[\theta, \theta] = r_1 \cdot m_2(\theta) + (1 - r_1) \cdot m_1(\theta) \cdot m_2(\theta) \qquad (10)$$

The values r_1 and r_2 are *overlapping factors* which measure the amount of overlapping in the evidence from the two sources. Intuitively, r_1 is the portion of $m_1(t)$ that relies upon overlapping evidence from $m_2(t)$. The assumption is that the amount of overlapping between two pieces of evidence will affect their inter-dependence.

Estimation of the Overlapping Factors. We provide semantics for the overlapping factors using the probability theory. The detailed formulation can be found in Appendix A. The definition of r_i requires knowing certain conditional probabilities ($Pr[w_{i\pm1}|w_i]$ in the Appendix), which is not available. Thus we need to estimate r_i just as we need to estimate the bpa's for the sensors. In SnIPS each alert node is associated with a set of IDS signatures that triggered it. We view these signatures as different sensors (Fig. 1). In our analysis the identities of the sensors that triggered an alert are propagated to the hypotheses it supports and further along the graph to other hypotheses it implies. Thus each hypothesis such as h_1 or h_2 is associated with the set of sensors whose alerts support it. Each sensor s has a quality metric δ_s as discussed in Sect. 3.2. Let R_1 and R_2 be the two sensor sets associated with the hypothesis h_1 and h_2 to be combined using formula 6, and $R = R_1 \cap R_2$. We use formulas 11 or 12 to estimate the overlapping between h_1 and h_2.

$$r_1 = \frac{\sum_{s \in R} \delta_s}{\sum_{s \in R_1} \delta_s} \quad , \quad r_2 = r_1 \cdot \alpha \quad , \quad \alpha \le 1, \qquad (11)$$

$$r_2 = \frac{\sum_{s \in R} \delta_s}{\sum_{s \in R_2} \delta_s} \quad , \quad r_1 = r_2 \cdot \alpha^{-1}, \quad \alpha > 1, \qquad (12)$$

where α is defined in (A1) and can be computed as:

$$\alpha = \frac{m_1(t) \cdot (1 - m_2(t))}{m_2(t) \cdot (1 - m_1(t))} \qquad (13)$$

That is, we gauge the overlapping between the two sources by dividing the weight of the overlapping part by the total weight of each source, where the weight is calculated as the sum of the sensor quality metrics. Depending on the value of α, we estimate one of r_1, r_2 and compute the other using α. The above estimation ensures that both r_1 and r_2 are within $[0, 1]$. Appendix A gives further intuition behind the overlapping factor.

3.4 Belief Calculation Algorithm

Typically the alert correlation graph returned by SnIPS is not fully connected but contains a number of correlation graph segments like the one shown in Fig. 1. The algorithm in general takes a set of correlation graph segments and calculates the belief value for each node on each graph. The graph segments are then sorted in descending order based on the maximum belief values for the sink nodes. To calculate the belief of the sink node, the algorithm propagates the quality metrics of each alert in the graph. The propagation will use the translation relation between the semantics of nodes. The algorithm applies the extended combination rule when there are multiple arcs flowing into one node like node 9 in Fig. 1. IDS signature identifications are propagated throughout the graph to be used in estimating the overlapping factor using formulas 11 or 12. The complexity of this algorithm is *linear* in the size of the graph. In the worst case the SnIPS-generated graph is quadratic in the number of IP addresses in the alerts [29]. Appendix B has the formal algorithm with its details.

3.5 An Illustrative Example

We use the example in Fig. 1 to show the belief calculation process. It starts by computing the belief values for the source nodes alerts (node 1–5), each of which is associated with the sensor (IDS signature) that triggered it as in Sect. 3.2. Then the belief values will be propagated through the graph using the semantics of the source node to the destination node using a set of predefined translation (compatibility) tables. Combination will be needed when multiple derivation paths lead to a single node. Let us take node 9 as an example, which has three pieces of evidence flowing into it from node 6, 7, 8. All the parent nodes 6, 7, 8's belief values based on their perspective semantics are translated into the bpa on node 9's semantics ($compromised(ip_2)$). The algorithm sorts the three branches based on the translated belief values and combines the highest belief pair. In the similar manner the combined branches are further combined with the rest branches. Let us assume that node 7 and 8 are the first pair to combine. Node 7's belief value after translation is $m_1(t) = 0.68$ and node 8's value is $m_2(t) = 0.6$. First we need to estimate the correlation factors r_1 and r_2 using formulas (11) or (12). Let R1 and R2 be the two sensor sets for node 7 and 8. $R_1 = \{sensor_2, sensor_3\}$ and $R_2 = \{sensor_3, sensor_4\}$. The quality metrics for the sensors are $\delta_{sensor_2} = 0.2$, $\delta_{sensor_3} = 0.6$, and $\delta_{sensor_4} = 0.01$. Using formula 13, we have $\alpha = 1.42$. After using formula 12, since $\alpha > 1$, we have $r_2 = 0.98, r_1 = 0.69$. Then after applying rules 7–10, we get Table 2.

Table 2. Combination example

(h_1,h_2)	$\psi(h_1,h_2)$	h	(h_1,h_2)	$\psi(h_1,h_2)$	h
(t,t)	0.596	{t}	(t, θ)	0.084	{t}
(θ, t)	0.004	{t}	(θ, θ)	0.316	$\{\theta\}$

Finally, $Bel(\{true\}) = 0.68$, calculated by summing up all the subsets of $\{true\}$ values. The final step is to combine this result with the belief from node 6, which will be in a similar manner. The sensor set associated with the combined belief will be the union of the sensor sets from all branches.

3.6 A Note on Methodology

Absence of Evidence. Our current model just counts for the supporting evidence when they are present. On the other hand, DS can handle the absence of evidence as negative evidence by assigning weight to false in the computation table. Below we discuss the reason why we did not include this functionality.

Besides the well-known poor priors problem in Bayesian inference, there is a second challenge in real-world intrusion detection systems that needs to be addressed as well, which is that we typically have a very poor understanding of all the ways in which an attack occurs. IDS systems such as Snort are signature-oriented in that they are designed to detect the occurrence of a specific event or series of events that serve as an sensor for an underlying attack. By Bayesian methodology we should be able to measure and use in predictive analysis both the true positive and true negative rates of detection, when available. However, because these rates are measured in the laboratory under different conditions than the real environment traffic the claimed rates tend to be estimates of the real rates whose quality is undetermined. From systems administrators' experience we have learned that for signature-based systems true positive rates (i.e. $Prob(attack\ has\ occurred|alert\ has\ fired)$) are usually close to accurate whereas false negative rates (i.e. ($Prob(no\ attack\ has\ occurred|alert\ has\ not\ fired)$) are less so. The positive case is intuitive – the specificity of the signature in an alert leads us to believe that the attack under question may have occurred. (Yet, true positive rates are never 1 because multiple system behaviors, some of them unknown to us, may satisfy the same signature leading to false positives.) For the negative case, when a (possibly expected) signature event is not seen that may be either because the attack has not occurred (a true negative) or because the attack has occurred in an undetectable manner (a false negative). This is not symmetric with the positive case because in the negative case we are modeling attack behavior rather than system behavior. In keeping with our approach of making minimal assumptions about attacks, our belief strength is currently built on the true positive rate alone. In future work we can consider the use of the true negative rate for specific sensors that can reliably detect the absence of an attack. DS theory can handle this type of negative evidence with the corresponding compatability relations among positive and negative evidence defined. The computation formula will also need to be extended to handle mixed positive/negative evidence.

4 Experimental Results

We implemented the algorithms in Java, and have been applying the system on our departmental network with about 200 servers and workstations (including Windows, Linux, and Mac OS X). The Snort alert collection, correlation, and DS algorithm application were all carried out on a Ubuntu server running a Linux kernel version 2.6.32 with 16 GB of RAM on an eight-core Intel Xeon processor of CPU speed 3.16 GHz. So far we have not encountered any performance bottleneck in our algorithm.

4.1 Evaluation Methodology

The objective of our evaluation is to examine whether the belief values calculated from our DS algorithm can help a security analyst to prioritize further investigation. To that end, we assigned to an IDS alert the belief value which is the highest belief of the hypotheses that supports. This can be easily calculated from the alert correlation graph through linear traversal. If IDS alerts with high belief values turn to be more likely true alerts than those with low belief values, it is an indication of the effectiveness of our approach.

Moreover, to show that it is indeed the application of Dempster-Shafer theory helps in alert prioritization, we compared the performance of our DS algorithm against that of the following alternative methods:

1. Using sensor quality metrics only. In this method, we simply use the sensor quality metrics assigned to each alert as described in Sect. 3.2 as an alert's belief value.
2. Using the maximum sensor-quality metric in a correlation graph as the belief value for all alerts in the graph.
3. Using the belief values calculated from the standard DS rule of combination, instead of from our customized DS.

All these methods assign a belief value to an IDS alert. A threshold value was chosen. Alerts with belief values above the threshold will be classified as true alerts, and those below the threshold will be classified as false alerts.

We used the truth files that included in the data set to determine which alerts are actually true alerts and which are actually false alerts. Then we compared this against the classification provided by the belief values. The key metrics in the classifier's performance are precision, recall (true positive), and false positive. As the belief-value threshold is changed, the classifier will obtain different operating points in terms of true positive and false positive. We draw receiver operating characteristic (ROC) curves for the four methods and compared their performance. ROC curve is a standard way to compare performance of IDS systems [3]. It shows the relationship between the detection rate (true positive) and false positive rate of a classifier.

$$\text{precision} = \frac{\text{\# true alerts above threshold}}{\text{\# alerts above threshold}}$$

$$\text{recall (true positive)} = \frac{\text{\# true alerts above threshold}}{\text{\# total true alerts}}$$

$$\text{false positive} = \frac{\text{\# false alerts above threshold}}{\text{\# total false alerts}}$$

4.2   The Rationale for Using Lincoln Lab Data-Set

The first source of data we used in our evaluation is Snort alerts from the CIS departmental network at Kansas State University. Due to the lack of ground truth in such data from the production network, we provide anecdotal experiences on the effectiveness of our algorithm. In addition, we tested our prototype on the MIT Lincoln Lab DARPA intrusion detection evaluation data set. Although the Lincoln Lab data set has been criticized in the literature [16,17], it still one of small number of usable publicly available data sets for IDS research. This is due to its well-documented ground truth and the existence of both background and attack traffic. We believe the limitation of the (LL) dataset will not significantly affect the validity of our evaluation for the following reasons.

1. Most of the identified problems in the (LL) dataset would affect anomaly-based detection [16] where one needs to use the data for both training and testing purposes. These defects will not affect as much signature based IDS such as Snort, which we use as the underlying alert source.
2. Our reasoning model is built a priori from existing Snort rule repositories, and calibrated on our departmental network, completely unrelated to the (LL) data.
3. The problem in (LL) dataset's background traffic [17] makes it hard to make claims on the performance of the tested system on real networks. This is especially the case since it is a very old data set now. For this reason we will mainly use the dataset to compare performance. The relative performance of the various methods is likely not affected as much as the absolute performance, since they may all benefit or suffer from the specific features of the data set.

4.3   Lincoln Lab DARPA Data-Set Results

DARPA 1998 and 1999 Training Data. We obtained the training data[1] in packet capture (pcap) format for both the 1998 and 1999 DARPA Intrusion Detection Evaluation program. We ran Snort on the packet capture data, ran SnIPS on the alerts triggered by Snort, and ran our DS calculation algorithm as well as the other three methods mentioned in Sect. 4.1 on the generated alerts and correlation graphs. We created ground truth about alerts using the truth files provided at the data set website. Each day has attacks targeted at specific machines, as given in the truth files. We carefully went through each attack described and checked against the alert database to pick out those alerts that can be verified as true alerts. The rest of the alerts are false alerts. This ground truth allows us to calculate the true positive and false positive of the various classifiers and plot their corresponding ROC curves.

---

[1] Only training data's truth file is publicly available.

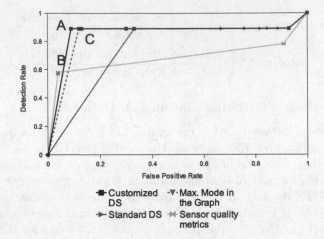

Fig. 2. Lincoln lab 1998 ROC curves

In general the steeper and closer the ROC curve is to the left-up corner, the better the classifier. A comparison of the ROC curves generated for both data sets is shown in Figs. 2 and 3. From the curves it is clear that our customized DS algorithm outperforms the other three alternative methods. Some operating points of the other three methods come close to the customized DS algorithm for the (LL 98) data, *e.g.* point B and C. But these points become much more inferior for the (LL 99) data. Whereas our DS algorithm produces the most optimal operating point consistently for both graphs (point A, corresponding to belief threshold 0.9).

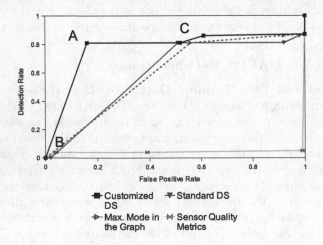

Fig. 3. Lincoln lab 1999 ROC curves

4.4   Sensitivity Analysis

We also did experiments to test how the variation in the choice of sensor quality metric values for the certainty tags affect our algorithm's performance. We varied the default mapping shown in Table 1 in four different ways, each of which perturbs the numeric value by about 10 %, e.g. from 0.33 to 0.3. We compared the results from all the four cases along with the default case in the ROC curves for the (LL 99) data (Fig. 4). One can find that the five curves exactly overlap with each other indicating that small perturbation in the values for the certainty tags has virtually no effect at all on the performance of the classifier. We did the same experiment for (LL 98) data and also obtained five overlapping curves.

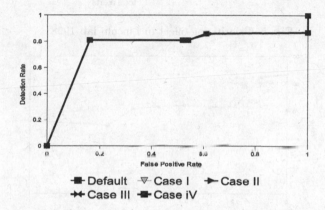

Fig. 4. Lincoln lab 1999 sensitivity analysis's ROC curves

4.5   Prioritization Effect

Our main objective of applying Dempster-Shafer theory is to use the relative belief values to prioritize intrusion analysis. Figures 5 and 6 show how the precision and recall changes when the threshold decreases from 1 to 0 (note that 0 in the X axis corresponds to belief 1, and 1 corresponds to belief 0). When one starts with alerts with high beliefs, the precision is high meaning more of the effort is devoted to useful tasks. When the threshold decreases, the cumulative precision decreases as well. This is a strong indication that the calculated belief values can be used effectively to prioritize further investigation.

At the highest belief range (0 point at the X axis) the percentage of total alerts captured is about 40 %, and the recall is about 80 %. This means that if one only analyzes alerts with the highest belief (e.g., >0.9), it only includes 40 % of all alerts whereas covers 80 % of all the true alerts. The recall curve is very flat meaning that most of the attacks can be captured using a high threshold value. This is certainly only the case for these two specific data sets, but nevertheless it indicates the effectiveness of prioritization provided by the DS method. Without it, one would have to look at twice as many alerts to achieve the same coverage.

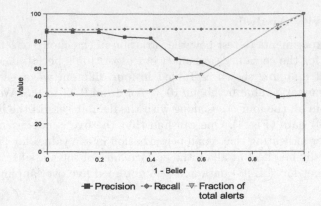

Fig. 5. Prioritizing effect in Lincoln lab 1998

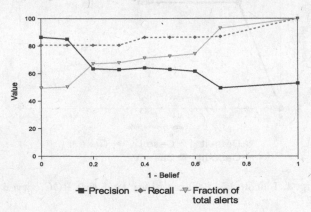

Fig. 6. Prioritizing effect in Lincoln lab 1999

4.6 The Production Network Results

In our evaluation we had permission from the University IT Security team
to monitor the Snort alerts from the Computing and Information Sciences
departmental network. This process had little privacy concern because the
links between the machines' *IPs* and the users were not given to us and Snort
only reveals limited payload data. All researchers involved in the experiments
had signed the University IT Personnel Ethics agreement, consistent with the
University policy.

Since it is hard to get ground truth in live systems, we presented the results
to the system administrator of our departmental network to get his feedback on
the tool's effectiveness. Regarding the quantitative belief calculation, the system administrator found that although the numbers themselves were hard to
interpret intuitively, the ranking would be useful in prioritizing further analysis.
He agreed that the higher-ranked correlations are indeed what he would like
to investigate further, compared with the lower-ranked ones. In most cases the

investigation indicated false positive or turned out inconclusive. But the ranking reduces the search scope for the system administrator which in real terms may translate to many man-hours of intrusion analysis oversight by a human. Certainly such anecdotal experiences cannot serve as validation of the method's effectiveness but real-world feedback is valuable in judging whether the tool is likely to be useful in the future.

5   Related Work

Chen, et al. [7] described the general approach of applying standard DS theory to combine multiple sensors' reports for intrusion detection in ad-hoc networks. *Yu, et al.* [34] extended Dempster-Shafer theory for alert fusion in the HPCN IDS alert correlation systems [33]. They observed that direct application of Dempster-Shafer theory in IDS alert fusion provides non-intuitive results and extended DS to weight alerts based on their quality. Our approach is different in that we directly capture sensor quality by assigning the remaining bpa to the unknown case ({*true, false*}), instead of the {*false*} case. We feel that our approach better captures the intuitive semantics provided by an IDS alert (Sect. 3.2). Neither *Chen* nor *Yu* addresses non-independence among evidence sources, which we believe is an important issue and have designed a customized DS combination rule to handle (Sect. 3.3).

There have also been approaches for alert fusion and prioritization based on decision theories. *Barreno, et al.* [4] introduce an optimal approach for combining binary classifiers using the Neyman-Pearson lemma. *Guofei, et al.* [12] propose an alert fusion technique based on likelihood ratio test (LRT). We would like to investigate the possibility that these techniques could be applied in an IDS alert correlation framework and compare the result with that of our DS-based approach.

Ou, et al. proposed an empirical approach to handling uncertainty in intrusion analysis [21]. They proposed using discrete tags to capture alert uncertainty in correlation analysis and a "proof-strengthening" technique to elevate confidence in a hypothesis where there are multiple derivation paths pointing to the same conclusion. The proof-strengthening rule is based on empirical experience and the authors did not provide the rationale behind it. Our approach takes discrete input metrics, but uses a quantitative combination method which provides a finer-grained result that can be used to rank hypotheses. Our quantitative approach has a well-established theoretical foundation, and can potentially provide better prioritization.

There have also been work on using Bayesian Network in intrusion detection [2] and IDS alert correlation [18,36]. The advantage of applying DS as opposed to Bayesian theory is that one does not need to know all the prior probabilities of events which are often unavailable. Indeed, DS is one of the various so-called non-traditional theories for uncertainty that generalize specific probabilities to an interval of probability, which also include Belief Theory, Subjective

Logic, and Possibility Theory. Some of these other approaches have been proposed in IDS alert fusion [30]. According to [22] the DS Belief Function theory is superior to the other theories because of

- the relatively high degree of theoretical development in DS theory,
- the aspect of Dempster-Shafer theory as a generalization of traditional probability theory, namely, where probabilities are assigned to sets of events as opposed to mutually exclusive singletons events,
- the versatility of DS theory in combining different types of evidence from multiple sources, and
- the large number of applications of DS theory in engineering in the past ten years.

IDS alert correlation [8,9,19,20,27,31–33,35,37] has been extensively studied in the literature. However, just because a correlation exists does not automatically mean the associated alerts are high confidence. The correlation itself are often "false correlations". From our conversation with system administrators, it is highly desirable that alert correlation tools prioritize their output based on the likelihood of true attacks. Our work provides one possible approach to this prioritization.

Denceux's work [10] explicitly raises and addresses the question of non-independent sources in DS theory. They point out that lack of independence in evidence is a valid concern in many applications and propose a new rule of combination called the "cautious rule" to handle this case. The cautious rule is designed to be as general as possible and is hence very complex and unintuitive. Our combination rule follows the general idea proposed by Shafer [26] and is based on a simple probabilistic semantics. It could be that our rule can be considered a highly specialized case of the general cautious rule, appropriate to our application.

Sun et al. [28] present an application of DS theory to the risk analysis of information systems security. They present an evidential reasoning approach that provides a rigorous, structured model to incorporate relevant risk factors, related counter measures and their interrelationships when estimating information system risk. *Chen et al.* [6] present an application of DS to the detection of anomalies in a variety of systems such as worm detection in email and learning in biological data. They show that by combining multiple (independent) signal sources using belief values and the Dempster combination rule, it is possible to achieve better results (characterized by rate of classification error) than by using a single signal. They point out that the advantage of using DS theory over Bayesian is that no *a priori* knowledge is required, making it potentially suitable for anomaly detection of previously unseen information whereas Bayesian inference requires *a priori* knowledge and does not allow allocating probability to ignorance.

6   Future Work

We will continue to apply our system on more production systems for extended periods of time, and gather data to further analyze its performance on real systems nowadays. There are more types of information than IDS alerts that could be incorporated into intrusion analysis; and Dempster-Shafer theory could be useful to reason about a much wider variety of dependency among various types of sensors, including non-monotonic dependencies. There are also other aspects such as temporal relationship that could affect the dependency. We plan to investigate along these directions when we gain more empirical experience of the method's effectiveness on production systems.

7   Conclusion

In this paper we presented a practical approach to prioritizing intrusion analysis using an extended Dempster-Shafer theory. The proposed DS application can correctly combine non-independent evidence commonly found in correlated IDS alerts. We proposed a DS model for capturing sensor quality that corresponds to the intuitive interpretation, and designed an algorithm for calculating belief values for hypotheses on an alert correlation graph. The main goal of this work is to reduce the workload on the system administrator by picking out those intrusion alerts that are most likely to be true and hence worthy of further investigation. We conducted rigorous evaluation of our approach on both a production network and two additional data sets. The results of evaluation strongly indicate that the ranking provided by the DS belief value gives good and robust prioritization on correlated alerts based on their likelihood of being true attacks. We believe our proposed approach will provide valuable practical tools to assist security analysts.

Acknowledgments. This material is based upon work supported by U.S. National Science Foundation under grant no. 1622402 and 1314925, AFOSR under Award No. FA9550-09-1-0138, and HP Labs Innovation Research Program. Any opinions, findings and conclusions or recommendations expressed in this material are those of the authors and do not necessarily reflect the views of the National Science Foundation, AFOSR, or Hewlett-Packard Development Company, L.P.

A   Semantics of the Overlapping Factors

Since we only have two non-zero bpa subsets: t and θ, in each hypothesis's frame of discernment, we use w_i to denote the fact that we trust h_i ($h_i = t$) and $\bar{w}_i$ (negation of w_i) to denote the fact that we do not trust h_i ($h_i = \theta$). One may find it strange that w_i and $\bar{w}_i$ appear to be not mutually exclusive, since θ includes both t and f. This is exactly the unique way in which DS expresses disbelief in a hypothesis – it differentiates clearly between not believing a hypothesis and believing the negation of that hypothesis. When we trust a hypothesis, we believe its state is t and when we do not trust a hypothesis, we do not know what its state is, hence θ. Interested readers are referred to Shafer's discussion on how to handle non-independent evidence using this interpretation [26]. The semantics of overlapping factor can be defined as:

$$r_1 = \frac{Pr[w_2|w_1] - Pr[w_2]}{Pr[\bar{w_2}]}, r_2 = \frac{Pr[w_1|w_2] - Pr[w_1]}{Pr[\bar{w_1}]}$$

Let us take r_1 as an example to explain the semantics. If we condition on trusting hypothesis h_1, the probability that we also trust h_2 is greater than or equal to its absolute probability since shared IDS sensors only give us positive correlation. The bigger the difference, the stronger influence trusting h_1 has on trusting h_2. The extreme case is when $Pr[w_2|w_1] = 1$, which gives $r_1 = 1$. Both r_1 and r_2 measure the dependence between w_1 and w_2, but from different directions.

Theorem A1

$$r_2 = \alpha \cdot r_1, \ where \ \alpha = \frac{Pr[w_1] \cdot Pr[\bar{w_2}]}{Pr[w_2] \cdot Pr[\bar{w_1}]} \tag{14}$$

Proof.

$$r_1 \cdot Pr[\bar{w_2}] \cdot Pr[w_1] = Pr[w_1, w_2] - Pr[w_1] \cdot Pr[w_2]$$
$$r_2 \cdot Pr[\bar{w_1}] \cdot Pr[w_2] = Pr[w_1, w_2] - Pr[w_1] \cdot Pr[w_2]$$

We then have

$$r_1 \cdot Pr[\bar{w_2}] \cdot Pr[w_1] = r_2 \cdot Pr[\bar{w_1}] \cdot Pr[w_2]$$

Theorem A2

$$\psi[h_1, h_2] = Pr[w_1, w_2]$$

Proof. Let us substitute r_i's into formulas (7)–(10). Let us also substitute the following definitions:

$$m_i(t) = Pr[w_i] \ m_i(\theta) = Pr[\bar{w_i}]$$

knowing that:

$$Pr[w_2|w_1] = \frac{Pr[w_1, w_2]}{Pr[w_1]}.$$

then substitute the above into the definition of r_1, we get

$$r_1 \cdot Pr[\bar{w_2}] \cdot Pr[w_1] = Pr[w_1, w_2] - Pr[w_1] \cdot Pr[w_2]$$

knowing that $Pr[\bar{w_2}] = 1 - Pr[w_2]$, then:

$$Pr[w_1, w_2] = r_1 \cdot Pr[w_1] + (1 - r_1) \cdot Pr[w_1] \cdot Pr[w_2]$$
$$= \psi[t, t]$$

The importance of this theorem is that our way of calculating the joint bpa $\psi[h_1, h_2]$ is sound in that it gives a generalization of the joint probability distribution of the trustworthiness of two (potentially) dependent sources. This also follows Shafer's general guide on how to handle non-independent evidence sources in DS [26], although Shafer did not provide the specific formulations.

B   Belief Calculation Algorithm

The main algorithm is **DsCorr** (Algorithm 1). This function takes *GraphSet* which is a set of correlation graph segments. It iterates on each graph, and returns a set of the graph segments sorted by the belief of the sink node (or highest sink node for multiple sinks) in descending order.

Algorithm 1. Rank graph segments by belief

```
1: function DsCorr(GraphSet)
2:      for each Graph in GraphSet do
3:          MakeAcyclic(Graph)
4:          ProcessingQueue ← all the source nodes
5:          while (ProcessingQueue is not empty) do
6:              Node ProcessingQueue.RemoveHead
7:              ComputeNodeBelief(Node)
8:              Node.visited ← true
9:              for each c in Node.Children do
10:                     if all c's parents are marked visited
11:                         AND c is not visited then
12:                             add c into ProcessingQueue
13:                     end if
14:                 end for
15:             end while
16:             record the highest belief value of sink nodes.
17:     end for
18:     return SortGraphSetbyBelief(GraphSet)
19: end function
```

Algorithm **ComputeNodeBelief** (Algorithm 2) takes a node and returns the belief value of it. There are three cases to consider for the node: (1) it is a source node; (2) it has only one parent node, (3) it has multiple parents. In the first case **AssignBpaValues** is called to compute the basic probability assignment based on the method in Sect. 3.2. This case applies to the alert nodes, *e.g.*, node 1–5 in Fig. 1. In the second case the node has only one parent so the translation function is called. The third case for combination is done by first translating implicitly into the node and then combine.

Algorithm 2. Compute the belief of a node

```
 1: function COMPUTENODEBELIEF(Node)
 2:     if Node has no parents then
 3:         ASSIGNBPAVALUES(Node)
 4:     else if Node has one parent p then
 5:         Node.belief ← TRANSLATE(p)
 6:         Node.sigSet ← p.sigSet
 7:     else if Node has multiple parents ps then
 8:         Node.belief ← COMBINE(ps)
 9:         Node.sigSet ← union of ps.sigSet
10:     end if
11: end function
```

References

1. Snort rules documentation. http://www.snort.org
2. Almgren, M., Lindqvist, U., Jonsson, E.: A multi-sensor model to improve automated attack detection. In: 11th International Symposium on Recent Advances in Intrusion Detection (RAID 2008). RAID, September 2008
3. Axelsson, S.: The base-rate fallacy and the difficulty of intrusion detection. ACM Trans. Inf. Syst. Secur. **3**(3), 186–205 (2000)
4. Barreno, M., Cárdenas, A.A., Tygar, J.D.: Optimal ROC curve for a combination of classifiers. In: Advances in Neural Information Processing Systems (NIPS, 2007) (2008)
5. Carrier, B.: A hypothesis-based approach to digital forensic investigations. Technical report, Center for Education and Research in Information Assurance and Security (CERIAS), Purdue University (2006)
6. Chen, Q., Aickelin, U.: Anomaly detection using the Dempster-Shafer method. In: International Conference on Data Mining (DMIN 2006) (2006)
7. Chen, T.M., Venkataramanan, V.: Dempster-Shafer theory for intrusion detection in ad hoc networks. IEEE Internet Comput. **9**, 35–41 (2005)
8. Cheung, S., Lindqvist, U., Fong, M.W.: Modeling multistep cyber attacks for scenario recognition. In: DARPA Information Survivability Conference and Exposition (DISCEX III), Washington, D.C., pp. 284–292 (2003)
9. Cuppens, F., Miège, A.: Alert correlation in a cooperative intrusion detection framework. In: IEEE Symposium on Security and Privacy (2002)
10. Denceux, T.: The cautious rule of combination for belief functions and some extensions. In: 9th International Conference on Information Fusion (2006)
11. Fine, T.L.: Theories of Probability. Academic Press, New York (1973)
12. Guofei, G., Cárdenas, A.A., Lee, W.: Principled reasoning and practical applications of alert fusion in intrusion detection systems. In: Proceedings of the 2008 ACM Symposium on Information, Computer and Communications Security, ASIACCS 2008, pp. 136–147. ACM, New York (2008)
13. Halpern, J.Y.: Reasoning About Uncertainty. The MIT Press, London (2005)
14. Jensen, F.V., Nielsen, T.D.: Bayesian Networks and Decision Graphs. Springer, New York (2007)
15. ArgusLab.: Snort intrusion analysis using proof strengthening (SnIPS). http://people.cis.ksu.edu/xou/argus/software/snips/

16. Mahoney, M.V., Chan, P.K.: An analysis of the 1999 DARPA/Lincoln laboratory evaluation data for network anomaly detection. In: Proceedings of the Sixth International Symposium on Recent Advances in Intrusion Detection (RAID) (2003)
17. McHugh, J.: Testing intrusion detection systems: a critique of the 1998 and 1999 DARPA intrusion detection system evaluations as performed by Lincoln laboratory. ACM Trans. Inf. Syst. Secur. (TISSEC) 3(4), 262–294 (2000)
18. Modelo-Howard, G., Bagchi, S., Lebanon, G.: Determining placement of intrusion detectors for a distributed application through Bayesian network modeling. In: 11th International Symposium on Recent Advances in Intrusion Detection (RAID 2008). RAID, September 2008
19. Ning, P., Cui, Y., Reeves, D., Dingbang, X.: Tools and techniques for analyzing intrusion alerts. ACM Trans. Inf. Syst. Secur. 7(2), 273–318 (2004)
20. Noel, S., Robertson, E., Jajodia, S.: Correlating intrusion events and building attack scenarios through attack graph distances. In: 20th Annual Computer Security Applications Conference (ACSAC 2004), pp. 350–359 (2004)
21. Xinming, O., Raj Rajagopalan, S., Sakthivelmurugan, S.: An empirical approach to modeling uncertainty in intrusion analysis. In: Annual Computer Security Applications Conference (ACSAC), December 2009
22. Sentz, K., Ferson, S.: Combination of evidence in Dempster-Shafer theory. Technical report, Sandia National Laboratories, Albuquerque, New Mexico (2002)
23. Shafer, G.: A Mathematical Theory of Evidence. Princeton University Press, Princeton (1976)
24. Shafer, G.: The problem of dependent evidence. Technical report, University of Kansas (1984)
25. Shafer, G.: Belief functions and possibility measures. In: The Analysis of Fuzzy Information (1986)
26. Shafer, G.: Probability judgment in artificial intelligence and expert systems. Stat. Sci. 2(1), 3–16 (1987)
27. Smith, R., Japkowicz, N., Dondo, M., Mason, P.: Using unsupervised learning for network alert correlation. In: Bergler, S. (ed.) Canadian AI. LNCS (LNAI), vol. 5032, pp. 308–319. Springer, Heidelberg (2008)
28. Sun, L., Srivastava, R.P., Mock, T.J.: An information systems security risk assessment model under Dempster-Shafer theory of belief functions. J. Manag. Inf. 22, 109–142 (2006)
29. Sundaramurthy, S.C., Zomlot, L., Xinming, O.: Practical IDS alert correlation in the face of dynamic threats. In: The 2011 International Conference on Security and Management (SAM 2011), Las Vegas, USA, July 2011
30. Svensson, H., Audun Jøsang.: Correlation of intrusion alarms with subjective logic. In: Sixth Nordic Workshop on Secure IT systems (NordSec) (2001)
31. Valeur, F.: Real-time intrusion detection alert correlation. Ph.D. thesis, University of California, Santa Barbara, May 2006
32. Valeur, F., Vigna, G., Kruegel, C., Kemmerer, R.A.: A comprehensive approach to intrusion detection alert correlation. IEEE Trans. Dependable Secure Comput. 1(3), 146–169 (2004)
33. Dong, Y., Frincke, D.: A novel framework for alert correlation and understanding. In: International Conference on Applied Cryptography and Network Security (ACNS) (2004)
34. Dong, Y., Frincke, D.: Alert confidence fusion in intrusion detection systems with extended Dempster-Shafer theory. In: 43rd ACM Southeast Conference, Kennesaw, GA, USA (2005)

35. Zhai, Y., Ning, P., Xu, J.: Integrating IDS alert correlation and OS-level dependency tracking. In: Mehrotra, S., Zeng, D.D., Chen, H., Thuraisingham, B., Wang, F.-Y. (eds.) ISI 2006. LNCS, vol. 3975, pp. 272–284. Springer, Heidelberg (2006)
36. Zhai, Y., Ning, P., Iyer, P., Reeves, D.S.: Reasoning about complementary intrusion evidence. In: Proceedings of 20th Annual Computer Security Applications Conference (ACSAC), pp. 39–48, December 2004
37. Zhou, J., Heckman, M., Reynolds, B., Carlson, A., Bishop, M.: Modeling network intrusion detection alerts for correlation. ACM Trans. Inf. Syst. Secur. (TISSEC) **10**(1), 4 (2007)
38. Zomlot, L., Sundaramurthy, S.C., Luo, K., Xinming, O., Raj Rajagopalan, S.: Prioritizing intrusion analysis using Dempster-Shafer theory. In: 4TH ACM Workshop on Artificial Intelligence and Security (AISec) (2011)

On the Soundness of Attack Trees

Maxime Audinot[✉] and Sophie Pinchinat[✉]

IRISA, Campus de Beaulieu, 35042 Rennes Cedex, France
{maxime.audinot,sophie.pinchinat}@irisa.fr

Abstract. We formally define three notions of soundness of an attack tree w.r.t. the system it refers to: *admissibility, consistency,* and *completeness.* The system is modeled as a labeled transition system and the attack is provided with semantics in terms of paths of the transition system. We show complexity results on the three notions of soundness, and the influence of the operators that are in the attack tree (see the recap in Fig. 5).

1 Introduction

Attack trees [4,5,8] are graphical representations of sets of attacks described in a hierarchical manner. The hierarchy is reflected in the structure of the tree, whose internal nodes represent abstract attack goals, and leaf nodes represent atomic goals. Internal nodes of an attack tree have extra information, namely the combinator (or operator) which expresses how the goal of the current node decomposes into the combination of its children goals. Classic operators are the "or" operator, the "sequential" operator, and the "and" operator.

Attack trees are a common tool used in risk analysis. The tree is used to describe the attacks to which of a system is vulnerable. First, an attack tree is constructed from a model of the system, and then it is analyzed for quantitative results, like computing the likelihood of an attack. In this paper, we focus on the qualitative part of attack trees, because our trees can be post-processed to take likelihood into account by adding weights to the leafs and propagating them.

There are different ways of defining the semantics of attack trees, which unsurprisingly strongly relies on the semantics of the set of operators. In [5], the focus is put on quantitative interpretations: atomic goals are given values in a domain, then, via the operators' semantics, a bottom-up computation yields a value at the (root node of the) tree that corresponds to, *e.g.* the length of the shortest attack, the highest probability to achieve an attack, etc.

In this contribution, we propose various semantics of attack trees that enable us to interpret them in the context of the system they refer to. This is strongly motivated by the nature of the top-down manual design of attack trees by practitioners, where the leaves a tree are iteratively refined into a combination of sub-nodes. To our knowledge, this issue has not been addressed in the literature.

In our setting, the system the tree refers to is a standard transition system S labeled over a set of atomic propositions Prop. It represents the operational

© Springer International Publishing AG 2016
B. Kordy et al. (Eds.): GraMSec 2016, LNCS 9987, pp. 25–38, 2016.
DOI: 10.1007/978-3-319-46263-9_2

semantics of some domain, as done in [7] for military buildings, or in [6] for socio-technical systems, leaving aside quantitative aspects (likelihood, time, cost). We describe the *attack goal* of a node by an expression $\iota \blacktriangleright \gamma$, where $\iota, \gamma \in \mathrm{Prop}$ are atomic propositions that denote respectively the preconditions and post-conditions of the goal (in the spirit of automated planning approaches). A natural system-based denotational path semantics is given to an attack goal $\iota \blacktriangleright \gamma$, where ι and γ are atomic propositions: the denoted set of paths is composed of all paths of the fixed transition system $\mathcal{S}$ that start from a state labeled by the precondition ι and that end in a state labeled by the post-condition γ. The internal nodes of an attack tree carry an attack goal, together with the operator that describes its decomposition into sub-goals[1], hence a pair $(\iota \blacktriangleright \gamma, \square)$; we call such an internal node a $\square$-node. In this paper, we let $\square$ range over $\{\varovee, \varolessthan, \varowedge\}$ for the "or", the "sequential and", and the "and" operators respectively. In our graphical representations of attack trees (see Fig. 2 on Page 8), the shapes of the nodes emphasize the operator associated to the node: $\varovee$-nodes are represented with an ellipse, $\varolessthan$-nodes are represented with pentagons pointing rightwards, and $\varowedge$-nodes are represented with rectangles, and the leaf nodes are represented with rounded corners rectangles.

In this paper, we address the soundness of an attack tree in terms of the relationship between an internal node $(\iota \blacktriangleright \gamma, \square)$ and the list of its children nodes $(\iota_1 \blacktriangleright \gamma_1, \square_1), \ldots, (\iota_n \blacktriangleright \gamma_n, \square_n)$ (from left to right). To do so, we compare[2] the set of paths denoted by $\iota \blacktriangleright \gamma$ with the $\square$-combination of the sets of paths denoted by the children $\iota_i \blacktriangleright \gamma_i$ of that node.

We introduce three notions of soundness for attack trees w.r.t. the transition system: *admissibility*, *consistency*, and *completeness*. Admissibility captures the approach where practitioners decompose the main goal into a structured goal some of whom achievements are also an achievements of the main goal. Consistency expresses that the proposed decomposition of the main goal guarantees its achievement. Finally, the intent of completeness is a complete characterization of the main goal in terms of the proposed decomposition.

The three notions of soundness are defined by comparing the two sets of paths denoted by $\iota \blacktriangleright \gamma$ and the $\square$-combination of the sets of paths denoted by the children. We use the three natural comparisons between sets, namely equality, inclusion, and non-empty intersection. Each notion of soundness entails a decision problem, of whether a given attack tree is sound or not w.r.t. the transition system it refers to. We establish complexity results on the three notions of soundness, and with regards to the kinds of operators that are allowed. We show that the admissibility problem is in P for the operators $\varovee$ and $\varolessthan$, but becomes NP-complete for the operator $\varowedge$. Next, we prove that the consistency problem is in P for the operators $\varovee$, CO-NP[3] for the operator $\varolessthan$ and CO-NP-complete for the operator $\varowedge$. The completeness problem is in CO-NP for the operators

---

[1] The children of the internal node.

[2] See further for details.

[3] That is the *negative instances* of the decision problem, *i.e.* those for which the answer is "no", are fully characterized by a polynomial-time non-deterministic algorithm.

$\emptyset$ and $\ominus$, and in Π_2^P for the operator $\oslash$. Recall that Π_2^P is a complexity class of the polynomial hierarchy [10] composed of languages whose complement is in Σ_2^P, or equivalently NP^{NP}, that are languages captured by a non-determinitsic polynomial-time algorithm which can call a non-determinitsic polynomial-time subroutine[4].

The paper is organized as follows: In Sect. 2, we present preliminaries notions used in the rest of the paper. In Sect. 3, we present transitions systems and formal attack goals, and their paths properties. In Sect. 4, we present attack trees, and the three soundness completeness, consistency and admissibility. In Sect. 5, we show the complexity results for the three soundness. In Sect. 6, we discuss the complexity result and conjecture about the harness that are not established yet.

2 Preliminaries

For $i, j \in \mathbb{N}$, we denote by $[i; j]$ the *interval* of integers ranging over $\{i, i+1, \ldots j\}$. For a finite set X, 2^X is the powerset of X, $|X|$ is the cardinal of X, X^* is the set of finite sequences of elements of X. For a binary relation R over a set X ($R \subseteq X \times X$), we say that R is *left-total* if for every $x \in X$, there exists $y \in X$ such that $(x, y) \subset R$. We denote by R^* the reflexive and transitive closure of R.

We recall that P is the class of decision problems[5] that can be solved by a deterministic polynomial-time algorithm, that NP is the class of decision problems that can be solved by a non-deterministic polynomial-time algorithm, and CO-NP is the class of decision problems whose complementary problem[6] is in NP. As a typical representative of the class NP, we will consider the classical decision problem SAT (We refer to [3] for these classic classes of complexity). We end with the class Π_2^P of the polynomial hierarchy which captures the decision problems whose negative instances can be characterized by a non-determinitsic polynomial-time algorithm which can call a non-determinitsic polynomial-time subroutine[7]. We refer to [10] for details on the polynomial hierarchy.

3 Transition Systems and Attack Goals

In this section, we define transition systems, attack goals and the semantics of the operators $\{\emptyset, \ominus, \oslash\}$.

3.1 Transition Systems

Without loss of generality and for technical reasons, transition systems will carry no actions, but instead have all the necessary information in their states via a labeling by atomic propositions.

---

[4] Which is classically called an *oracle*.
[5] The answer is "Yes/No".
[6] The answers "Yes/No" are swapped.
[7] Which is classically called an *oracle*.

Definition 1 (Transition system). *Let* Prop *be a finite set of atomic propositions. A transition system* over Prop *is a tuple* $\mathcal{S} = (S, \rightarrow, \lambda)$, *where:*

- *S is the finite set of* states,
- $\rightarrow \subseteq S \times S$ *is the* transition relation *of the system (which is assumed left-total[8]),*
- $\lambda : \text{Prop} \rightarrow 2^S$ *is the* valuation *function.*

The size of $\mathcal{S}$ *is* $|\mathcal{S}| = |S| + |\rightarrow|$.

Let $S' \subseteq S$ *be a sub-set of states. We let* $Post^*_{\mathcal{S}}(S')$ *be the set of states that are reachable from some state of* S', *and* $Pre^*_{\mathcal{S}}(S')$ *be the set of states that are co-reachable from some state of* S'. *Formally,*

- $Post^*_{\mathcal{S}}(S') := \{s \in S \mid$ *there is some* $s' \in S$ *such that* $s' \rightarrow^* s\}$
- $Pre^*_{\mathcal{S}}(S') := \{s \in S \mid$ *there is some* $s' \in S$ *such that* $s \rightarrow^* s'\}$.

We will use the following running example:

Example 1. The set Prop_e is $\{i, f, m_1, m_2, e_1, e_2, pre_a, post_a, pre_b, post_b, pre_c, post_c\}$, the system $\mathcal{S}_e = (S_e, \rightarrow_e, \lambda_e)$ over Prop_e, whose graphical representation is given in Fig. 1, is formally defined by: $S_e = \{s_i\}_{0 \leq i \leq 6}$, where the transition relation $\rightarrow_e$ contains the pairs (s_0, s_1), (s_0, s_2), (s_1, s_3), (s_2, s_3), (s_2, s_4), (s_3, s_5), (s_4, s_6). Finally, we let $\lambda_e(i) = \lambda_e(pre_a) = \{s_0\}$, $\lambda_e(f) = \{s_5, s_6\}$, $\lambda_e(m_1) = \lambda_e(post_a) = \{s_1, s_2\}$, $\lambda_e(m_2) = \{s_3, s_4\}$, $\lambda_e(e_1) = \{s_5\}$, $\lambda_e(e_2) = \{s_6\}$, $\lambda_e(pre_b) = \{s_1, s_2, s_4\}$, $\lambda_e(post_b) = \{s_3, s_6\}$, $\lambda_e(pre_c) = \{s_2, s_3\}$, and finally, $\lambda_e(post_c) = \{s_4, s_5\}$. Also, $Pre^*_{\mathcal{S}}(\{s_3\}) = \{s_0, s_1, s_2, s_3\}$ and $Post^*_{\mathcal{S}_e}(\{s_1, s_6\}) = \{s_1, s_3, s_5, s_6\}$.

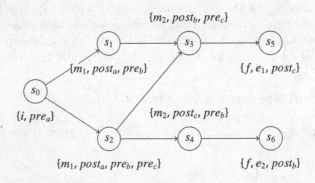

Fig. 1. Example of transition system: $\mathcal{S}_e$.

Definition 2 (Paths, elementary paths, factors). *A* path *in a system* $\mathcal{S}$ *is a sequence of states of the form* $\pi = s_0 s_1 \ldots s_n \in S^*$ *for some* $n \in \mathbb{N}$, *such that for all* $k \in [0; n-1], (s_k, s_{k+1}) \in \rightarrow$. *An* elementary path *is a path* $s_0 s_1 \ldots s_n$ *where* $\forall k \neq k' \in [0; n], s_k \neq s_{k'}$ *(i.e. there is no cycles). We denote by* $\Pi(\mathcal{S})$ *the*

---

[8] This is classic and it is no loss of generality.

set of paths of $\mathcal{S}$. Let $\pi = s_0 \ldots s_n \in \Pi(\mathcal{S})$. The length *of π is n, written $|\pi|$. A* factor *of π is a sequence $s_i \ldots s_j$ for some $0 \leq i \leq j \leq n$, that will be denoted by $\pi[i;j]$. The interval $[i;j]$ is an* anchoring interval, *or simply an* anchoring, *of the factor $\pi[i;j]$ in π.*

We define two notions of decomposition of a path that reflect the refinement of attack tree nodes. Both are based on factors.

Definition 3 (Sequential and parallel decomposition of paths). *Let $\pi \in \Pi(\mathcal{S})$ be a path. A sequence $\pi_1 \ldots \pi_k \in \Pi(\mathcal{S})^*$ of paths is a* sequential decomposition *of π if each π_j is a factor of π, there are ordered anchorings of the π_j's that form a tiling of the interval $[0; |\pi|]$. In particular, this anchoring of π_1 is of the form $[0; y]$ and this anchoring of π_k is of the form $[x; |\pi|]$. A set of paths $\{\pi_1, \ldots, \pi_k\}$ is a* parallel decomposition *of $\pi \in \Pi(\mathcal{S})$ if each π_j is a factor of π, and the anchorings of the paths π_j cover the interval $[0; |\pi|]$. Notice that a sequential decomposition is a particular case of a parallel decomposition.*

Example 2. Consider the path $\pi = s_0 s_2 s_3 s_5$ in system $\mathcal{S}_e$. The sequence $(s_0 s_2).(s_2 s_3 s_5)$ is a sequential decomposition of π, where the anchoring of $s_0 s_2$ is (unique and equal to) $[0; 1]$ and the unique anchoring of $s_2 s_3 s_5$ is $[1; 3]$. The set $\{s_2 s_3 s_5, s_0 s_2 s_3\}$ is a parallel decomposition of π, where the anchoring of $s_2 s_3 s_5$ is $[1; 3]$ and the anchoring of $s_0 s_2 s_3$ is $[0; 2]$.

For the Sects 3 2 and 4, we fix a transition system $\mathcal{S} = (S, \rightarrow, \lambda)$.

3.2   Attack Goals

Attack goals are descibed in a formal language meant to specify attack objectives that internal nodes of an attack tree naturally carry.

Definition 4 (Attack goals). *An* attack goal *is an expression of the form either $\iota \blacktriangleright \gamma$, or a term of the form $(\iota_1 \blacktriangleright \gamma_1) \square \ldots (\iota_n \blacktriangleright \gamma_n)$, where $\square \in \{\oslash, \ominus, \oslash\!\!\!\!\triangle\}$ and $\iota, \iota_1, \ldots \iota_n, \gamma, \gamma_1, \ldots \gamma_n \in \text{Prop}$.*

Example 3. $i \blacktriangleright f$, $(i \blacktriangleright e_1) \oslash (i \blacktriangleright e_2)$ and $(i \blacktriangleright post_a) \ominus (post_a \blacktriangleright post_c) \ominus (post_c \blacktriangleright post_b)$ are attack goals, whose interpretation will be given in system $\mathcal{S}_e$ (see Example 4).

Definition 5 (Path semantics of attack goals). *The* path semantics *of an attack goal t, written $[t]_{\mathcal{S}}^{path}$, is a subset of $\Pi(\mathcal{S})$ defined by: if $t = \iota \blacktriangleright \gamma$, then $[\iota \blacktriangleright \gamma]_{\mathcal{S}}^{path} = \{\pi \in \Pi(\mathcal{S}) \mid \pi(0) \in \lambda(\iota) \text{ and } \pi(|\pi|) \in \lambda(\gamma)\}$, otherwise we distinguish between the different operators $\square \in \{\oslash, \ominus, \oslash\!\!\!\!\triangle\}$ according to:*

$$[(\iota_1 \blacktriangleright \gamma_1) \oslash_n (\iota_2 \blacktriangleright \gamma_2) \oslash_n \ldots (\iota_n \blacktriangleright \gamma_n)]_{\mathcal{S}}^{path} = \bigcup_{1 \leq i \leq n} [\iota_i \blacktriangleright \gamma_i]_{\mathcal{S}}^{path}$$

$$[(\iota_1 \blacktriangleright \gamma_1) \ominus_n (\iota_2 \blacktriangleright \gamma_2) \ominus_n \ldots (\iota_n \blacktriangleright \gamma_n)]_{\mathcal{S}}^{path} = \{\pi \mid \text{there is a}$$
decomposition $\pi_1.\pi_2 \ldots .\pi_n$ of π and each $\pi_i \in [\iota_i \blacktriangleright \gamma_i]_{\mathcal{S}}^{path}\}$

$$[(\iota_1 \blacktriangleright \gamma_1) \oslash\!\!\!\!\triangle_n (\iota_2 \blacktriangleright \gamma_2) \oslash\!\!\!\!\triangle_n \ldots (\iota_n \blacktriangleright \gamma_n)]_{\mathcal{S}}^{path} = \{\pi \mid \text{there is a}$$
parallel decomposition $\{\pi_1, \pi_2, \ldots, \pi_n\}$ of π and each $\pi_i \in [\iota_i \blacktriangleright \gamma_i]_{\mathcal{S}}^{path}\}$

Example 4. The attack goals of Example 3 have the following path semantics:

$$[i \blacktriangleright f]_{\mathcal{S}_e}^{\text{path}} = \{\pi \in \Pi(\mathcal{S}_e) \mid \pi(0) = s_0 \text{ and } \pi(|\pi|) \in \{s_5, s_6\}\}$$
$$[(i \blacktriangleright e_1) \oslash_2 (i \blacktriangleright e_2)]_{\mathcal{S}_e}^{\text{path}} = [i \blacktriangleright f]_{\mathcal{S}_e}^{\text{path}}$$
$$[(i \blacktriangleright post_a) \ominus_3 (post_a \blacktriangleright post_c) \ominus_3 (post_c \blacktriangleright post_b)]_{\mathcal{S}_e}^{\text{path}} = \{s_0 s_2 s_4 s_6\}$$

4   Attack Trees

We define the set of attack trees over a set Prop of atomic propositions. In addition to the classical branching structure with nodes typed by a operator, we decorate each node with an attack goal $\iota \blacktriangleright \gamma$, representing the goal of the node. This goal is a formalization of the what is usually written in plain text in nodes of classical attack trees.

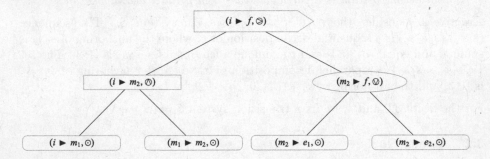

Fig. 2. The attack tree T_e.

An *attack tree* (over Prop) is either a leaf of the form $(\iota \blacktriangleright \gamma, \odot)$ or a composed tree of the form $(\iota \blacktriangleright \gamma, \square)(T_1, T_2 \ldots T_n)$, where $\iota, \gamma \in$ Prop, $\square \in \{\oslash, \ominus, \oslash\}$, $n \geq 2$, and $T_1, T_2, 1 \ldots, T_n$ are attack trees. We call the main node of a non-leaf tree a $\square$-*node* whenever it is of the form $(\iota \blacktriangleright \gamma, \square)(T_1, T_2 \ldots T_n)$.

The *path semantics* of an attack tree $(\iota \blacktriangleright \gamma, \square)(T_1, T_2 \ldots T_n)$ is naturally defined as $[\iota \blacktriangleright \gamma]_{\mathcal{S}}^{\text{path}} \subseteq \Pi(\mathcal{S})$.

Example 5. Figure 2 represents the attack tree T_e over Prop_e where:

$$T_e = (i \blacktriangleright f, \ominus)((i \blacktriangleright m_2, \oslash)((i \blacktriangleright m_1, \odot), (m_1 \blacktriangleright m_2, \odot)),$$
$$(m_2 \blacktriangleright f, \oslash)((m_2 \blacktriangleright e_1, \odot), (m_2 \blacktriangleright e_2, \odot)))$$

Another example of such an attack tree, using an $\oslash$-node, is $(i \blacktriangleright f, \oslash)(pre_a \blacktriangleright post_a, \odot), (pre_b \blacktriangleright post_b, \odot), (pre_c \blacktriangleright post_c, \odot))$.

For example, the path semantics of T_e is $\Pi(\mathcal{S}_e)$.

We now turn to more subtle semantics for attack trees that enable one to relate a tree with its subtrees, or equivalently an internal node with its children, in terms of their path semantics, hence the explicit reference to the system the attack tree refers to. Our proposal yields three notions of soundness with different interpretations from the point of view of the practitioner. The *admissibility* property means that there is an attack that achieves the parent node goal that decomposes with the ones of its children nodes (Eq. (1)). The *consistency* property means that the combined children node goals yield attacks (if any) that achieve the parent node goal (Eq. (2)). Finally, the *completeness* property means that the combined children node goals fully characterize the parent node goal (Eq. (3)).

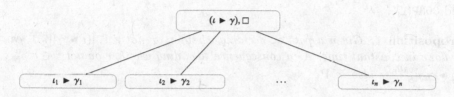

Fig. 3. A picture for Eq. (1).

Definition 6 (Admissibility). *The attack tree* $(\iota \blacktriangleright \gamma, \square)(T_1, T_2 \ldots T_n)$ *is admissible w.r.t.* $\mathcal{S}$ *either when* $\square$ *is* $\odot$, *or when Eq.* (1) *holds, where* $\iota_i \blacktriangleright \gamma_i$ *is the local attack goal of the tree* T_i $(1 \le i \le n)$, *see Fig. 3.*

$$[\, \square_{i=1}^n (\iota_i \blacktriangleright \gamma_i)]_{\mathcal{S}}^{path} \cap [\iota \blacktriangleright \gamma]_{\mathcal{S}}^{path} \ne \emptyset \tag{1}$$

Then, the *consistency* and *completeness* properties are variants of the admissibility property by replacing Eq. (1) of Definition 6 by Eqs. (3) and (2), respectively:

$$[\, \square_{i=1}^n (\iota_i \blacktriangleright \gamma_i)]_{\mathcal{S}}^{path} \subseteq [\iota \blacktriangleright \gamma]_{\mathcal{S}}^{path} \tag{2}$$

$$[\, \square_{i=1}^n (\iota_i \blacktriangleright \gamma_i)]_{\mathcal{S}}^{path} = [\iota \blacktriangleright \gamma]_{\mathcal{S}}^{path} \tag{3}$$

Remark 1. As Eq. (3) entails Eq. (2), completeness implies consistency.

For example, the attack tree $(i \blacktriangleright f, \oslash)((i \blacktriangleright m_2, \odot), (m_2 \blacktriangleright f, \odot))$ is admissible w.r.t. $\mathcal{S}_e$, whereas $(pre_a \blacktriangleright post_a, \oslash)((pre_b \blacktriangleright post_b, \odot), (pre_c \blacktriangleright post_c, \odot))$ is not admissible w.r.t. $\mathcal{S}_e$.

5 The Decision Problems $\text{ADM}(\mathcal{O}), \text{CONS}(\mathcal{O}), \text{COMP}(\mathcal{O})$

We formalize the decision problems $\text{ADM}(\mathcal{O}), \text{CONS}(\mathcal{O}), \text{COMP}(\mathcal{O})$ respectively related to the notions of admissibility, consistency and completeness, as introduced in Sect. 4. Let $\mathcal{O} \subseteq \{\oslash, \oslash, \oslash\}$.

Definition 7. *The* Admissibility *problem* $\text{ADM}(\mathcal{O})$ *is defined by:*
Input: $\theta = \iota \gamma \iota_1 \gamma_1 \ldots \iota_n \gamma_n$ *a sequence of atomic propositions,* $\square \in \mathcal{O}$ *and* $\mathcal{S}$ *a labeled transition system over* $\{\iota, \gamma, \iota_1, \ldots, \gamma_n\}$.
Output: *"yes" if* $(\iota \blacktriangleright \gamma, \square)((\iota_1 \blacktriangleright \gamma_1, \odot), (\iota_2 \blacktriangleright \gamma_2, \odot) \ldots (\iota_n \blacktriangleright \gamma_n, \odot))$ *is admissible w.r.t* $\mathcal{S}$, *"no" otherwise.*

We similarly define the decisions problems $\texttt{CONS}(\mathcal{O})$ and $\texttt{COMP}(\mathcal{O})$ in a natural way, respectively called the *the consistency problem* and *the completeness problem*. In the sequel, we will denote by $(\theta, \square, \mathcal{S})$ an instance of $\texttt{ADM}(\mathcal{O})$, $\texttt{CONS}(\mathcal{O})$, or $\texttt{COMP}(\mathcal{O})$, where unless explicitly stated, θ expands into $\iota \gamma \iota_1 \gamma_1 \ldots \iota_n \gamma_n$.

5.1 Preliminary Complexity Results

We first establish useful technical propositions that will be used to prove our main results on complexity for the three decision problems $\texttt{ADM}(\mathcal{O})$, $\texttt{CONS}(\mathcal{O})$, and $\texttt{COMP}(\mathcal{O})$.

Proposition 1. *Given a path π, deciding whether or not $\pi \in [(\iota \blacktriangleright \gamma)]_{\mathcal{S}}^{path}$ can be done in constant time. As a consequence, deciding whether or not $\pi \in [\oslash_{i=1}^{n} (\iota_i \blacktriangleright \gamma_i)]_{\mathcal{S}}^{path}$ is also in* P.

Proof. For $\pi \in [(\iota \blacktriangleright \gamma)]_{\mathcal{S}}^{path}$, the only thing to check is $\pi(0) \in \lambda(\iota)$ and $\pi(|\pi|) \in \lambda(\gamma)$. For $\pi \in [\oslash_{i=1}^{n} (\iota_i \blacktriangleright \gamma_i)]_{\mathcal{S}}^{path}$, it amount to finding $i \in [1; n]$ such that $\pi \in [(\iota_i \blacktriangleright \gamma_i)]_{\mathcal{S}}^{path}$.

The next two propositions address operators $\ominus$ and $\oslash$.

Proposition 2. *Given a path π, deciding whether or not $\pi \in [\ominus_{i=1}^{n} (\iota_i \blacktriangleright \gamma_i)]_{\mathcal{S}}^{path}$ is in* P.

Proof. It sufficient to check that $\pi(0) \in \lambda(\iota_1)$ and $\pi(|\pi|) \in \lambda(\gamma_n)$ and to make a traversal of π that seeks for a sequence of positions $0 \leq y_1 \leq \cdots \leq y_{n-1} \leq |\pi|$ such that $\pi(y_i) \in \lambda(\gamma_i) \cap \lambda(\iota_{i+1})$, for all $i \in [1; n-1]$.

Proposition 3. *Given a path π, deciding whether or not $\pi \in [\oslash_{i=1}^{n} (\iota_i \blacktriangleright \gamma_i)]_{\mathcal{S}}^{path}$ is in* NP.

Proof. To verify that $\pi \in [\oslash_{i=1}^{n} (\iota_i \blacktriangleright \gamma_i)]_{\mathcal{S}}^{path}$ the algorithm guesses n factors of π, or equivalently their sequence of anchorings $[x_1; y_1], \ldots, [x_n; y_n]$ and checks they provide a parallel decomposition of π. Namely, the algorithm needs to check the following properties: (i) $x_i \leq y_i$, for each $i \in [1; n]$, (ii) for each $x \in [0; |\pi|]$, there exists $i \in [1; n]$ such that $x_i \leq x \leq y_i$, and (iii) $\pi[x_i; y_i] \in [\iota_i \blacktriangleright \gamma_i]_{\mathcal{S}}^{path}$, that is $\pi(x_i) \in \lambda(\iota_i)$ and $\pi(y_i) \in \lambda(\gamma_i)$. By the above propositions, it is clear that Properties (i)-(iii) can be verified in polynomial time.

The two following propositions are helpful in order to bound the size of the paths we will need to guess in our non-deterministic algorithms of Sect. 5.2.

Proposition 4. *Let $\square \in \{\oslash, \ominus, \oslash\}$. If $[\square_{i=1}^{n} (\iota_i \blacktriangleright \gamma_i)]_{\mathcal{S}}^{path} \neq \emptyset$, then it contains a path of size smaller than $|S| (2n - 1)$. In particular, if $n = 1$, we can consider a path of length at most $|S|$, that is an elementary path.*

Proof. We first consider the case where $\square = \oslash$. Let $\pi \in [\oslash_{i=1}^n (\iota_i \blacktriangleright \gamma_i)]_S^{\text{path}}$, and let $[x_1; y_1], \ldots, [x_n; y_n]$ be the anchoring intervals of a parallel decomposition of π, such that, for each $1 \leq i \leq n$, $\pi(x_i) \in \lambda(\iota_i)$ and $\pi(y_i) \in \lambda(\gamma_i)$. Let $z_1 \leq \cdots \leq z_{2n}$ be the resulting of sorting the elements $x_1, y_1 \ldots, x_n, y_n$. Notice that the sequence $\pi[z_1; z_2], \pi[z_2; z_3], \ldots, \pi[z_{2n-1}; z_{2n}]$ is a sequential decomposition of π. For $1 \leq i \leq 2n - 1$, let π_i' be the elementary path obtained from $\pi[z_i; z_{i+1}]$ by removing the cycles. We have $|\pi_i'| \leq |S|$. The path π' obtained by the sequential composition of the paths π_i' is in $[\oslash_{i=1}^n (\iota_i \blacktriangleright \gamma_i)]_S^{\text{path}}$ since the states $\pi(z_i)$ for $i \in [1; 2n]$ are still in π' and in the same order. Then we have $|\pi'| \leq |S|(2n - 1)$, which concludes. Regarding the case where $\square = \ominus$, it is enough to remark that sequential decomposition is a particular parallel decomposition, and the case $\square = \oslash$ is obvious as even elementary paths suffice.

Finally, the following more involved Proposition 5 plays a key role in our proofs of Sect. 5.2. The rest of this section is dedicated to its proof.

Proposition 5. *Given a $S = (S, \to, \lambda)$ be a labeled transition system over a set of propositions* Prop $\supseteq \{\iota, \gamma, \iota_1, \ldots, \gamma_n\}$, *it is* NP-*complete to decide whether or not $[\oslash_{i=1}^n (\iota_i \blacktriangleright \gamma_i)]_S^{\text{path}} \neq \emptyset$.*

NP-easyness: We describe the non-deterministic algorithm that decides $[\oslash_{i=1}^n (\iota_i \blacktriangleright \gamma_i)]_S^{\text{path}} \neq \emptyset$. This algorithm guesses a path π such that $|\pi| \leq |S|(2n - 1)$ (which is sufficient by Proposition 4), and n anchoring intervals $[x_1; y_1], \ldots, [x_n; y_n]$ in π. It then verifies that $\pi \in [\oslash_{i=1}^n (\iota_i \blacktriangleright \gamma_i)]_S^{\text{path}}$, which can be done in polynomial time in $(\theta, \oslash, S)$ according to Propositions 3.

NP-hardness: We reduce the classical NP-complete problem SAT [2] to $[\oslash_{i=1}^n (\iota_i \blacktriangleright \gamma_i)]_S^{\text{path}} \neq \emptyset$. An input of SAT is a set of clauses $\mathscr{C}$ over a set of propositions $\{p_1, \ldots p_r\}$, where each clause $C \in \mathscr{C}$ is a set of literals, that is either a proposition p_i or its negation $\neg p_i$. The SAT problem amounts to answering whether or not $\mathscr{C}$ is satisfiable, that is whether or not there is a valuation of the propositions $p_1, \ldots p_r$ that makes all clauses of $\mathscr{C}$ true. Now, let $\mathscr{C} = \{C_1, \ldots C_m\}$ over propositions $\{p_1, \ldots p_r\}$ be an input of the SAT problem; classically, $|\mathscr{C}| = \sum_{C \in \mathscr{C}} |C|$, where $|\mathscr{C}|$ is the number of literals that occur in C. We introduce two fresh propositions ι_0 and γ_0 and we define a labeled transition system $S_{\mathscr{C}} = (S_{\mathscr{C}}, \to_{\mathscr{C}}, \lambda_{\mathscr{C}})$ over Prop$_{\mathscr{C}} = \{\iota_0, \gamma_0, C_1, \ldots C_m\}$: In the following, we let ℓ_i denote either p_i or $\neg p_i$, for every $i \in \{1, \ldots, r\}$. We let $S_{\mathscr{C}} = \{s, t\} \cup \{\ell_i\}_{i=1,\ldots,r}$, $\to_{\mathscr{C}} = \{(\ell_i, \ell_{i+1}) \mid i \in [1; r - 1]\} \cup \{(s, \ell_1), (\ell_r, t)\}$, and $\lambda_{\mathscr{C}} : $ Prop $\to 2^S$ is such that $\lambda_{\mathscr{C}}(\iota_0) = \{s\}$, $\lambda_{\mathscr{C}}(\gamma_0) = \{t\}$, and $\lambda_{\mathscr{C}}(\ell) = C$ whenever ℓ is a literal of C. An example of $S_{\mathscr{C}}$ is depicted in Fig. 4. Notice that by definition, $|S_{\mathscr{C}}|$ is polynomial in $|\mathscr{C}|$.

In the following, let call a *full path* a path of $S_{\mathscr{C}}$ from s to t. The system $S_{\mathscr{C}}$ is designed in such a way that any full path visits either p_j or $\neg p_j$ in an exclusive manner, for each $i = 1, \ldots, r$. A full path π therefore unambiguously denotes a valuation v_π of the propositions. Reciprocally, every valuation v of the propositions yields a unique full path π_v. Additionally, a full path π visits a state

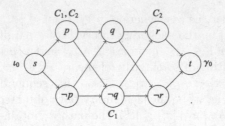

Fig. 4. $\mathcal{S}_{\mathscr{C}}$ with $\mathscr{C} = \{C_1, C_2\}$ where $C_1 = p \vee \neg q$ and $C_2 = p \vee r$

labeled by $C \in \mathscr{C}$ if, and only if, v_π makes the clause C true. Moreover, a full path π visits a state labeled by C if, and only if there is an anchoring $[0; j]$ such that $\pi(j)$ is labeled by C. The following concludes the proof of Proposition 5.

It remains to establish that $[(\iota_0 \blacktriangleright \gamma_0) \oslash (\iota_0 \blacktriangleright C_1) \oslash \ldots \oslash (\iota_0 \blacktriangleright C_m)]_{\mathcal{S}_\mathscr{C}}^{\text{path}} \neq \emptyset$ if, and only if, $\mathscr{C}$ is satisfiable.

Assume $[(\iota_0 \blacktriangleright \gamma_0) \oslash (\iota_0 \blacktriangleright C_1) \oslash \ldots \oslash (\iota_0 \blacktriangleright C_m)]_{\mathcal{S}_\mathscr{C}}^{\text{path}} \neq \emptyset$, with some element π. The constraint $(\iota_0 \blacktriangleright \gamma_0)$ enforces π to be a full path. Now, the other constraints enforce π to visits a state labeled by C for each $C \in \mathscr{C}$, entailing the valuation v_π making all clauses true, so that $\mathscr{C}$ is a positive input of SAT. It is not hard to see that conversely, if $\mathscr{C}$ is a positive instance of SAT, then any full path π_v, for some valuation v that makes all clauses true, is in $[(\iota_0 \blacktriangleright \gamma_0) \oslash (\iota_0 \blacktriangleright C_1) \oslash \ldots \oslash (\iota_0 \blacktriangleright C_m)]_{\mathcal{S}_\mathscr{C}}^{\text{path}}$.

5.2 Complexity Results for $\text{ADM}(\mathcal{O}), \text{CONS}(\mathcal{O}), \text{COMP}(\mathcal{O})$

We successively study admissibility, consistency, and completeness.

Theorem 1. $\text{ADM}(\{\oslash, \ominus\})$ *is in* P.

Proof. We consider the syntactic fragment of the temporal logic CTL [1] defined by: $\varphi ::= p \mid \varphi \wedge \varphi \mid \varphi \vee \varphi \mid \Diamond\varphi$, where the only temporal operator is the "eventually" one denoted by $\Diamond$. The semantics of a formula φ of this fragment is given with regard to a labeled transition system $\mathcal{S} = (S, \rightarrow, \lambda)$, and is noted $[\![\varphi]\!]_\mathcal{S}$. We define $[\![\varphi]\!]_\mathcal{S} \subseteq S$ by induction over φ: $[\![p]\!]_\mathcal{S} = \lambda(p)$, $[\![\varphi \wedge \varphi']\!]_\mathcal{S} = [\![\varphi]\!]_\mathcal{S} \cap [\![\varphi']\!]_\mathcal{S}$, $[\![\varphi \vee \varphi']\!]_\mathcal{S} = [\![\varphi]\!]_\mathcal{S} \cup [\![\varphi']\!]_\mathcal{S}$, and $[\![\Diamond\varphi]\!]_\mathcal{S} = Pre_\mathcal{S}^*([\![\varphi]\!]_\mathcal{S})$, where $Pre_\mathcal{S}^*$ is defined in Sect. 3.1, that is $s \in [\![\Diamond\varphi]\!]_\mathcal{S}$ iff there is a path in $\mathcal{S}$ starting from s that visits a state in $[\![\varphi]\!]_\mathcal{S}$. In the following we simply write $[\![\varphi]\!]$ instead of $[\![\varphi]\!]_\mathcal{S}$. Note that computing $[\![\varphi]\!]$ takes polynomial time in the size of $\mathcal{S}$ and φ, see for example [9].

Let $\varphi_\oslash^\theta := \bigvee_{i=1}^n \iota \wedge \iota_i \wedge \Diamond(\gamma \wedge \gamma_i)$. Then $[\oslash_{i=1}^n (\iota_i \blacktriangleright \gamma_i)]_\mathcal{S}^{\text{path}} \cap [\iota \blacktriangleright \gamma]_\mathcal{S}^{\text{path}} \neq \emptyset$ if, and only if, $[\![\varphi_\oslash^\theta]\!] \neq \emptyset$. Indeed, if $[\oslash_{i=1}^n (\iota_i \blacktriangleright \gamma_i)]_\mathcal{S}^{\text{path}} \cap [\iota \blacktriangleright \gamma]_\mathcal{S}^{\text{path}} \neq \emptyset$, then let π be a path in $[\oslash_{i=1}^n (\iota_i \blacktriangleright \gamma_i)]_\mathcal{S}^{\text{path}} \cap [\iota \blacktriangleright \gamma]_\mathcal{S}^{\text{path}}$. Then π go from ι to γ and there is an $i \in [1; n]$ such that π go from ι_i to γ_i. Hence, π go from $\iota \wedge \iota_i$ to $\gamma \wedge \gamma_i$. So $\pi \in [\![\iota \wedge \iota_i \wedge \Diamond(\gamma \wedge \gamma_i)]\!]$, which implies $\pi \in [\![\varphi_\oslash^\theta]\!]$, which is then non-empty.

Conversely, if $\llbracket \varphi^\theta_\oslash \rrbracket \neq \emptyset$, then let $\pi \in \llbracket \varphi^\theta_\oslash \rrbracket$. Then there is an $i \in [1; n]$ such that π start from ι_i and eventually reaches $\iota \wedge \gamma \wedge \gamma_i$. Let π' be the prefix of π that is a path that go from ι_i to $\iota \wedge \gamma \wedge \gamma_i$. We have $\pi' \in [\oslash_{i=1}^n (\iota_i \blacktriangleright \gamma_i)]_\mathcal{S}^{\text{path}} \cap [\iota \blacktriangleright \gamma]_\mathcal{S}^{\text{path}}$, so it is not empty.

Let $\varphi^\theta_\ominus := \iota \wedge \iota_1 \wedge \Diamond(\gamma_1 \wedge \iota_2 \wedge \Diamond(\gamma_2 \wedge \ldots \Diamond(\gamma_n \wedge \gamma)))$. Then $[\ominus_{i=1}^n (\iota_i \blacktriangleright \gamma_i)]_\mathcal{S}^{\text{path}} \cap [\iota \blacktriangleright \gamma]_\mathcal{S}^{\text{path}} \neq \emptyset$ if, and only if, $\llbracket \varphi^\theta_\ominus \rrbracket \neq \emptyset$ Indeed, if $[\ominus_{i=1}^n (\iota_i \blacktriangleright \gamma_i)]_\mathcal{S}^{\text{path}} \cap [\iota \blacktriangleright \gamma]_\mathcal{S}^{\text{path}} \neq \emptyset$, then let π be a path in $[\ominus_{i=1}^n (\iota_i \blacktriangleright \gamma_i)]_\mathcal{S}^{\text{path}} \cap [\iota \blacktriangleright \gamma]_\mathcal{S}^{\text{path}}$. Then π go from ι to γ and is the concatenation of paths π_i going from ι_i to γ_i. Hence, π visits successively $\iota \wedge \iota_1$, $\gamma_1 \wedge \iota_2$, $\ldots$ and $\gamma_n \wedge \gamma$. So $\pi \in \llbracket \varphi^\theta_\ominus \rrbracket$, which is then non-empty. Conversely, if $\llbracket \varphi^\theta_\ominus \rrbracket \neq \emptyset$, then let $\pi \in \llbracket \varphi^\theta_\ominus \rrbracket$. Then π visits successively $\iota \wedge \iota_1$, $\gamma_1 \wedge \iota_2$, $\ldots$ and $\gamma_n \wedge \gamma$. Let π' be the prefix of π that is a path that stops in $\gamma_n \wedge \gamma$. We have $\pi' \in [\ominus_{i=1}^n (\iota_i \blacktriangleright \gamma_i)]_\mathcal{S}^{\text{path}} \cap [\iota \blacktriangleright \gamma]_\mathcal{S}^{\text{path}}$, so it is not empty.

Theorem 2. ADM($\{\oslash\}$) *is* NP-*complete.*

Proof. First, we show that ADM($\oslash$) is NP-easy by giving a non-deterministic polynomial algorithm. According to Proposition 5, we can guess $\pi \in [\oslash_{i=1}^n (\iota_i \blacktriangleright \gamma_i)]_\mathcal{S}^{\text{path}}$ with a non-deterministic polynomial algorithm. Then, we check that $\pi \in [(\iota \blacktriangleright \gamma)]_\mathcal{S}^{\text{path}}$ by checking that $\pi(0) \in \lambda(\iota)$ and $\pi(|\pi|) \in \lambda(\gamma)$ which is done in constant time, so the algorithm is still polynomial.

Second, we prove the NP-hardness by reducing the problem of Proposition 5. Let $\mathcal{S} = (S, \rightarrow, \lambda)$ be a labeled transition system over propositions $\{\iota_1, \gamma_1, \ldots, \iota_n, \gamma_n\}$. We extend $\mathcal{S}$ to $\mathcal{S}' = (S, \rightarrow, \lambda')$ over propositions $\{\iota_1, \gamma_1, \ldots, \iota_n, \gamma_n\} \cup \{\iota, \gamma\}$, with $\lambda'(\iota) = \lambda'(\gamma) = S$ and λ' coincide with λ over $\{\iota_1, \gamma_1, \ldots, \iota_n, \gamma_n\}$. As $[(\iota \blacktriangleright \gamma)]_{\mathcal{S}'}^{\text{path}} = \Pi(\mathcal{S}')$, deciding ADM($\oslash$) for $\mathcal{S}$ amounts to deciding if $[\oslash_{i=1}^n (\iota_i \blacktriangleright \gamma_i)]_{\mathcal{S}'}^{\text{path}} \neq \emptyset$. So the reduction is straightforward, and ADM($\oslash$) is NP-hard.

Theorem 3. CONS($\{\oslash\}$) *is in* P.

Proof. We show a polynomial algorithm deciding CONS($\{\oslash\}$).
Let $(\theta, \oslash, \mathcal{S})$ be an input of CONS($\{\oslash\}$). First, we compute the sets of state $S(\iota_i) = \lambda(\iota_i) \cap Pre_\mathcal{S}^*(\lambda(\gamma_i))$ and $S(\gamma_i) = \lambda(\gamma_i) \cap Post_\mathcal{S}^*(\lambda(\iota_i))$ for $i \in [1; n]$, which is done in polynomial time by reachability analysis. $S(\iota_i)$ (resp. $S(\gamma_i)$) represents the states of $\lambda(\iota_i)$ (resp. $\lambda(\gamma_i)$) from which there is a path to a state of $\lambda(\gamma_i)$ (resp. from a state of $\lambda(\iota_i)$). Then, check that for every $i \in [1; n]$, $S(\iota_i) \subseteq \lambda(\iota)$ and $S(\gamma_i) \subseteq \lambda(\gamma)$.

Theorem 4. CONS($\{\ominus\}$) *is in* CO-NP.

Proof. We describe a polynomial non-deterministic algorithm deciding that an input $(\theta, \ominus, \mathcal{S})$ of CONS($\{\ominus\}$) is a negative instance[9]. Let $(\theta, \ominus, \mathcal{S})$ be an input of CONS($\{\ominus\}$). We first guess a path π, and check that $\pi \in [\ominus_{i=1}^n (\iota_i \blacktriangleright \gamma_i)]_\mathcal{S}^{\text{path}}$: the latter can be done by guessing anchoring intervals $[x_i; x_{i+1}]$ of a sequential decomposition of π, and by checking that for all $i \in [1; n]$, $x_i \leq x_{i+1}$, that

---

[9] Namely that the answer is "no".

$\pi(x_i) \in \lambda(\iota_i)$ and $\pi(x_{i+1}) \in \lambda(\gamma_i)$, and that $x_1 = 0$ and $x_{n+1} = |\pi|$. It then remains to verify that $\pi \notin [(\iota \blacktriangleright \gamma)]_{\mathcal{S}}^{\text{path}}$. Clearly all these checks can be done in polynomial time.

Theorem 5. $\text{CONS}(\{\oslash\})$ *is* CO-NP-*complete.*

Proof. A negative instance $(\theta, \oslash, \mathcal{S})$ of $\text{CONS}(\{\oslash\})$ is characterized by the existence of a path $\pi \in [\oslash_{i=1}^{n} (\iota_i \blacktriangleright \gamma_i)]_{\mathcal{S}}^{\text{path}} \setminus [(\iota \blacktriangleright \gamma)]_{\mathcal{S}}^{\text{path}}$.

Deciding that $(\theta, \oslash, \mathcal{S})$ is a negative instance of $\text{CONS}(\{\oslash\})$ is NP-easy: According to Proposition 5, we can guess $\pi \in [\oslash_{i=1}^{n} (\iota_i \blacktriangleright \gamma_i)]_{\mathcal{S}}^{\text{path}}$ with a non-deterministic polynomial-time algorithm. It then remains to check that $\pi \notin [(\iota \blacktriangleright \gamma)]_{\mathcal{S}}^{\text{path}}$ which takes constant time, by Proposition 1.

Deciding that $(\theta, \oslash, \mathcal{S})$ is a negative instance of $\text{CONS}(\{\oslash\})$ is NP-hard: we reduce the problem of deciding $[\oslash_{i=1}^{n} (\iota_i \blacktriangleright \gamma_i)]_{\mathcal{S}}^{\text{path}} \neq \emptyset$ (which is NP-complete by Proposition 5). Assume $\mathcal{S} = (S, \rightarrow, \lambda)$ is a labeled transition system over $\{\iota_1, \gamma_1, \ldots, \iota_n, \gamma_n\}$. We extend λ to the set of propositions $\{\iota_1, \gamma_1, \ldots, \iota_n, \gamma_n\} \cup \{\iota, \gamma\}$, by letting $\lambda(\iota) = \lambda(\gamma) = \emptyset$. By construction $[(\iota \blacktriangleright \gamma)]_{\mathcal{S}'}^{\text{path}} = \emptyset$, so that $(\theta, \oslash, \mathcal{S})$ is a negative instance of $\text{CONS}(\{\oslash\})$ is equivalent to $[\oslash_{i=1}^{n} (\iota_i \blacktriangleright \gamma_i)]_{\mathcal{S}'}^{\text{path}} \neq \emptyset$, and we are done.

We now turn to the decision problems $\text{COMP}(\mathcal{O})$. By Remark 1, a negative instance of $\text{CONS}(\mathcal{O})$ necessarily is a negative instance of $\text{COMP}(\mathcal{O})$.

Theorem 6. $\text{COMP}(\{\odot, \ominus\})$ *is in* CO-NP.

Proof. Let $(\theta, \square, \mathcal{S})$ be a negative instance of $\text{COMP}(\{\odot, \ominus\})$. There are two cases: (a) $(\theta, \square, \mathcal{S})$ is a negative instance of $\text{CONS}(\square)$, or (b) there exists a path $\pi \in [(\iota \blacktriangleright \gamma)]_{\mathcal{S}}^{\text{path}} \setminus [\square_{i=1}^{n} (\iota_i \blacktriangleright \gamma_i)]_{\mathcal{S}}^{\text{path}}$. The algorithm guesses whether it is Case (a) or Case (b). For Case (a) we use the polynomial-time algorithm in the proof of Theorem 3 (recall P $\subseteq$ CO-NP) for operator $\odot$ and Theorem 4 for operator $\ominus$. Regarding Case (b), we use a variant of the proof of Theorem 4: the non-deterministic polynomial-time algorithm that decides whether or not an input $(\theta, \square, \mathcal{S})$ is a negative instance of $\text{COMP}(\{\odot, \ominus\})$ consists in guessing a path π, and in checking that $\pi \in [(\iota \blacktriangleright \gamma)]_{\mathcal{S}}^{\text{path}} \setminus [\square_{i=1}^{n} (\iota_i \blacktriangleright \gamma_i)]_{\mathcal{S}}^{\text{path}}$. First, the algorithm guesses an elementary path π and checks $\pi \in [(\iota \blacktriangleright \gamma)]_{\mathcal{S}}^{\text{path}}$; this check is done in constant time by Proposition 1. Next, the algorithm checks that $\pi \notin [\square_{i=1}^{n} (\iota_i \blacktriangleright \gamma_i)]_{\mathcal{S}}^{\text{path}}$. If $\square = \odot$, we apply Proposition 1, otherwise $\square = \ominus$ and we use Proposition 2.

The last theorem shows that the operator $\oslash$ is much harder to handle.

Theorem 7. $\text{COMP}(\oslash)$ *is in* Π_2^P.

Proof. We show that negative instances of $\text{COMP}(\oslash)$ can be captured by a polynomial-time non-deterministic algorithm that can call a polynomial-time non-deterministic subroutine. Let $(\theta, \oslash, \mathcal{S})$ be a negative instance of $\text{COMP}(\oslash)$. Similarly to the case of Theorem 6, there are two possible cases: (a) $(\theta, \oslash, \mathcal{S})$ is a negative instance of $\text{CONS}(\oslash)$, or (b) there exists $\pi \in [(\iota \blacktriangleright \gamma)]_{\mathcal{S}}^{\text{path}} \setminus [\oslash_{i=1}^{n}$

$(\iota_i \blacktriangleright \gamma_i)]_{\mathcal{S}}^{\text{path}}$. Therefore, the algorithm first non-deterministically guesses if it is Case (a) or Case (b). For Case (a), it behaves like the polynomial-time non-deterministic algorithm proposed for $\text{CONS}(\oslash)$ in the proof of Theorem 5. Regarding Case (b), the algorithm guesses a path[10] π and checks that $\pi \in [(\iota \blacktriangleright \gamma)]_{\mathcal{S}}^{\text{path}}$. Then, the algorithm checks that $\pi \notin [\oslash_{i=1}^{n} (\iota_i \blacktriangleright \gamma_i)]_{\mathcal{S}}^{\text{path}}$. The latter can be performed by running an NP oracle according to Proposition 3. Hence the set of negative instances of $\text{COMP}(\oslash)$ is in NP^{NP}, that is Σ_2^P, which concludes.

	ADM	CONS	COMP
$\oslash$	P	P	co-NP
$\ominus$	P	co-NP	co-NP
$\oslash$	NP-complete	co-NP-complete	Π_2^P

Fig. 5. Complexities of the three decision problems for each operator

6   Discussion

In this paper, we have developed a path semantics for attack trees that yields three natural notions of soundness of attack trees: *admissibility*, *consistency*, and *completeness*. Each soundness notion conveys a meaning of the practioners' manual decomposition of internal nodes. We then have explored the complexity of the associated decision problems.

As can be seen, operators $\oslash$ and $\ominus$ are much simpler than the very classic operator $\oslash$ widely used in the literature; actually the complexity Π_2^P established here for the decision problem $\text{COMP}(\oslash)$ is not proved to be optimal, but we conjecture it is. This rather high complexity is not surprising since the notion of parallel decomposition underlying the operational semantics of operator $\oslash$ features complex combinatorics. As future work, we wish to complete the complexity classes picture by showing that, *e.g.* all complexities are tight.

References

1. Clarke, E.M., Emerson, A.E.: Design and synthesis of synchronization skeletons using branching time temporal logic. In: Kozen, D. (ed.) Logic of Programs. LNCS, vol. 131, pp. 52–71. Springer, Heidelberg (1981)
2. Cook, S.A.: The complexity of theorem-proving procedures. In: Conference Record of Third Annual ACM Symposium on Theory of Computing, Shaker Heights, Ohio, 3–5 May 1971, pp. 151–158 (1971)
3. Garey, M.R., Johnson, D.S.: Computers and Intractability. A Guide to the Theory of NP-Completeness. Freeman, New York (1979)

---

[10] By Proposition 4 it is enough to consider paths whose size is polynomial in $(\theta, \oslash, \mathcal{S})$.

4. Jhawar, R., Kordy, B., Mauw, S., Radomirović, S., Trujillo-Rasua, R.: Attack trees with sequential conjunction. In: Federrath, H., Gollmann, D., Chakravarthy, S.R. (eds.) SEC 2015. IFIP AICT, vol. 455, pp. 339–353. Springer, Heidelberg (2015). doi:10.1007/978-3-319-18467-8_23

5. Kordy, B., Mauw, S., Radomirović, S., Schweitzer, P.: Attack-defense trees. J. Logic Comput. **24**(1), 55–87 (2014)

6. Lenzini, G., Mauw, S., Ouchani, S.: Security analysis of socio-technical physical systems. Comput. Electr. Eng. **47**, 258–274 (2015)

7. Pinchinat, S., Acher, M., Vojtisek, D.: ATSyRa: an integrated environment for synthesizing attack trees. In: Mauw, S., Kordy, B., Jajodia, S. (eds.) GraMSec 2015. LNCS, vol. 9390, pp. 97–101. Springer, Heidelberg (2016). doi:10.1007/978-3-319-29968-6_7

8. Schneier, B.: Attack trees. Dr. Dobb's J. Softw. Tools **24**(12), 21–29 (1999)

9. Schnoebelen, P.: The complexity of temporal logic model checking. Adv. Modal Logic **4**(393–436), 35 (2002)

10. Stockmeyer, L.J.: The polynomial-time hierarchy. Theoret. Comput. Sci. **3**(1), 1–22 (1976)

The Right Tool for the Job: A Case for Common Input Scenarios for Security Assessment

Xinshu Dong[1](✉), Sumeet Jauhar[1], William G. Temple[1], Binbin Chen[1],
Zbigniew Kalbarczyk[2], William H. Sanders[2], Nils Ole Tippenhauer[3],
and David M. Nicol[2]

[1] Advanced Digital Sciences Center, Singapore, Singapore
{xinshu.dong,sumeet.j,William.T,binbin.chen}@adsc.com.sg
[2] University of Illinois at Urbana-Champaign, Champaign, IL, USA
kalbarcz@illinois.edu, whs@uiuc.edu, nicol@crhc.uiuc.edu
[3] Singapore University of Technology and Design, Singapore, Singapore
nils_tippenhauer@sutd.edu.sg

Abstract. Motivated by the practical importance of security assessment, researchers have developed numerous model-based methodologies. However, the diversity of different methodologies and tool designs makes it challenging to compare their respective strengths or integrate their results. To make it more conducive to incorporate them for practical assessment tasks, we believe it is critical to establish a common foundation of security assessment inputs to support different methodologies and tools. As the initial effort, this paper presents an open repository of Common Input Scenarios for Security Assessment (CISSA) for different model-based security assessment tools. By proposing a CISSA design framework and constructing six initial scenarios based on real-world incidents, we experimentally show how CISSA can provide new insights and concrete reference points to both security practitioners and tool developers. We have hosted CISSA on a publicly available website, and envision that community effort in building CISSA would significantly advance the scientific and practical values of model-based security assessment.

1 Introduction

Understanding the system security level against cyber threats is critical for today's IT or IT-enabled infrastructures, such as cloud storage and computing services, banking and payment systems, or cyber-physical systems such as smart grids. Industries today adopt various compliance standards (e.g., NERC CIP [17]) to exercise best practices in assessing their systems' level of security. Despite promoting security in general, such compliance practices often do not sufficiently capture the inherent relationships among different security-related aspects of the studied infrastructure. To provide deeper insight and more rigorous assessment into the overall security of the infrastructures, recent years have witnessed a surge of interest in model-based security assessment. In model-based security assessment, the various aspects of systems, threats, security measures, and more importantly, how these aspects interact with each other,

© Springer International Publishing AG 2016
B. Kordy et al. (Eds.): GraMSec 2016, LNCS 9987, pp. 39–61, 2016.
DOI: 10.1007/978-3-319-46263-9_3

are abstracted into models, often through rigorously defined formalisms. Such methodologies then use these underlying models to evaluate — often quantitatively via some forms of calculi — the security level of a system against specified cyber threats [11, 24, 29].

Notable examples of model-based security assessment methodologies include the attack tree [25] and its enhancements (e.g., Boolean logic Driven Markov Process [23] and Attack-Defense Tree [10]), the attack graph [15, 21] and its embodiments (e.g., [18, 26]), Unified Modeling Language (UML)-based formalisms (e.g., Cyber Security Modeling Language [27]), and Petri-net based formalisms (e.g., ADversary VIew Security Evaluation [14]). By applying techniques from this impressive range of methodologies, security practitioners could potentially assess the security of their systems in a systematic, rigorous, and holistic manner.

However, these methodologies manifest great diversity in selecting the information on system details and adversaries that are used as input, in representing and processing them (i.e., the formalisms, the calculi, the tool designs etc.), as well as in the aspects of security assessment (e.g., security metrics) they produce as output [29]. It is therefore difficult to understand their different strengths, compare their results, or integrate them in meaningful ways to present a multifaceted assessment of the systems. We have faced these challenges developing our own model-based security assessment methodology [5, 30], and we believe that industry practitioners seeking to adopt security modeling tools face similar dilemmas in vetting and selecting methodologies.

In this paper, we propose the creation of a set of *common input scenarios for security assessment* (CISSA). As illustrated in Fig. 1, a rich and open repository of input scenarios would allow both the security practitioners and the research community to better compare different methodologies, which in turn will drive future research and development to address various real-world security assessment needs and challenges. Our work propose a specification framework for defining CISSA and provide six sample scenarios based on real-world attacks against various IT and IT-enabled systems. We use the samples to analyze the real-world security assessment needs and evaluate three representative model-based assessment tools accordingly. Ultimately, we hope that our initial efforts

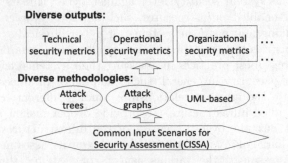

Fig. 1. Security assessment process driven by CISSA

will encourage the community to build a rich repository of CISSA for many different systems. Such a repository can be referenced by both tool developers and security practitioners in discussing the needs and features of security assessment. In summary, we make the following contributions in this paper:

- We propose a first blueprint design (CISSA) for specifying common input scenarios for different model-based security assessment tools.
- To initiate the effort, we construct six CISSA cases based on real-world incidents, which are hosted on a public CISSA repository [1]. Such example cases show the feasibility of building a diverse, realistic, structured, and precise set of CISSA based on the framework.
- We experimentally demonstrate the potential usefulness of CISSA by leveraging the six CISSA cases to investigate real-world needs of security assessment. We test three representative model-based security assessment tools and demonstrate that CISSA could provide new insights and concrete references to both security practitioners and tool developers.

In the rest of the paper, we propose our vision in Sect. 2, and then describe the elements of common input scenarios in Sect. 3. Section 4 reports our experiences in constructing six input scenarios, while a more detailed illustration of representing a real-world incident with the CISSA specification is presented in Appendix A. Section 5 uses the CISSA cases to investigate the security assessment needs and to test three representative security assessment tools. We discuss the roadmap for further developing CISSA in Sect. 6 and conclude in Sect. 7.

2   CISSA Vision

Devising effective means to assess the security of complex systems is one of the greatest challenges in security research [24,29], but the reward for solving this problem is profound. The complexity and diversity inherent in today's enterprise IT systems and critical infrastructures make it important to leverage a model-based approach to answer various security assessment questions, e.g.: How to design more resilient systems? How to make better investment decisions for different defense mechanisms? To realize the potential of model-based security assessment methodologies, we believe it is necessary to establish widely-accepted open input scenarios. Such common input scenarios will allow researchers and industry stakeholders to better understand the data required for security modeling, as well as the trade-offs and blind spots inherent in picking a tool or formalism for their system.

Such common input scenarios need to include the inputs required by security assessment tools based on individual real-world incidents or synthesized attacks. For example, many tools require the network topology and system configurations, as well as attacker capabilities as inputs to their assessment. Other tools may require specific additional information as inputs, such as the details of security countermeasures, and the estimated response time when suspicious behaviors are detected. The common input scenarios should be as comprehensive as possible in covering the general classes of inputs required by major security assessment tools. We envision that:

- For security practitioners, a common set of realistic CISSA cases will help them to conduct meaningful comparison and integration among different methodologies. Since assessment methods are validated on common, openly available, and realistic inputs, CISSA can help reduce the barrier to adoption for industry.
- For researchers and tool developers, a realistic, rich, and heterogeneous set of CISSA will make it clear which aspects to model and the important questions to answer. This will help guide the further development of the methodologies to meet real needs.
- An open repository of CISSA cases will allow any interested parties to contribute new cases that will be available to all. Overall, it will benefit the entire community in tool development and security assessment with greater interactions, synergy, and standard practices, enabling bigger impact of model-based security assessment for real-world systems.

Of course, realizing this vision will require a concerted effort on the part of the security assessment community. By proposing a blueprint for designing CISSA cases, hosting an open CISSA repository with six input scenarios as initial examples, we demonstrate the potential feasibility and usefulness of CISSA, hence encouraging other security researchers and tool developers to share the input scenarios they use.

3   Elements of an Input Scenario

Conceptually, an input scenario collects together the representation of the important security-relevant information surrounding a system. For industry practitioners, the scenario representation should allow them to describe the system they are responsible for and to express their concerns and design choices in cybersecurity aspects. For academics, the scenario representation will provide a reasonable proxy for real systems, to enable research and development of security assessment tools and methodologies.

We make the following choices when designing an initial blueprint for the CISSA framework:

- *Methodology-independent:* We decouple the raw security-related inputs/facts from any specific assessment methodologies. This would ensure that the input scenarios could be generally applied to study different security methodologies.
- *Comprehensive:* We want CISSA to cover different types of information, as long as it is security-related, i.e., the inclusion of the information can potentially affect the security assessment results. For example, CISSA should not only include the details of the attack, but elaborate on the corresponding environment (e.g., certain network topology, or specific software configuration) where the attack can occur as well. Because CISSA is not bound to any specific methodology, it is easy to define and further extend the framework to cover additional aspects that arise from the different kinds of systems/scenarios.

– *Realistic:* To help bridge the gap between academia and industry security
  practitioners, we believe CISSA should be made as realistic as possible. One
  way to ensure a high level of realism is to require CISSA to be based on real-
  world systems and security incidents. We expect the development of a CISSA
  uses the best information one can gather from the field.
– *Precise:* The information included in CISSA should represent different aspects
  of information in a structured way, and strive for minimum ambiguity, so as
  to enable objective and fair comparison among the tools.

Note that there are implicit conflicts and trade-offs among the different goals
we strive for, e.g., in order to make a case cover comprehensive information,
one often needs to include unstructured data; on the other hand, to ensure
that the included information is grounded, realistic, and precise, one often needs
to exclude hypothetical or unsubstantiated information, hence sacrificing the
comprehensiveness.

Guided by these design considerations, we propose a schema for representing security-relevant information. While there may be other ways to represent
such information, we believe that our schema (Fig. 2) could serve as a valuable
starting point for the CISSA concept. Existing resources such as security incident
reports are largely unstructured, and it would require human comprehension and
transformation before they can be used by security assessment tools. Databases
such as NVD provide machine-readable format, but only include information on
the specifics of the vulnerabilities, which is far from sufficient for most model-
based security assessment tools. In the design of CISSA, the identified elements
and their attributes provide a unified way to represent information on both the
target system and the security incident.

As shown in Fig. 2, a common input scenario for security assessment consists
of seven core elements, which are interrelated: system *components and network*,
data, *users*, and *operations*, as well as *undesirable outcomes*, *attacks*, and *countermeasures*. We represent these elements as a 7-tuple:

$$< N, D, U, O, X, A, C >.$$

We now elaborate on the characteristics of each element.

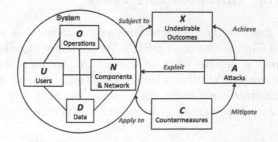

Fig. 2. Elements in an input scenario and their relationships

System Descriptions. The first four elements describe the system (in its general definition) to be assessed: N specifies the *components and network*, including both the specifications of the devices in the system (hardware/software, their configurations, and vulnerabilities), as well as the inter-connections and trust among them; D specifies the *data* that the system produces, stores, or consumes; U specifies the key *users* that interact with the system; and O specifies the *routine operations* that the system carries out. These four elements are tightly coupled: take the operation O as an example, it is defined in respect to N (e.g., which device is being controlled by a command, and which connection transmits the command), D (e.g., which sensor readings are used in an operation), and U (e.g., who takes charge of which step of the operation).

Undesirable Outcomes. Given the above system description, X specifies the *undesirable outcomes*, i.e., the security concerns, of the system. X is determined by the nature of the system and reflects its security requirements, including but not limited to the traditional Confidentiality, Integrity, and Availability triples. For example, for a railway system, engine failure, train crash, and fare evasion are some of the undesirable outcomes that can be potentially caused by attacks.

Attack. The element A describes the attack aspect, including both the concrete attack steps taken by an attacker and the information about the attacker itself: $A := <\alpha, \Sigma>$, where α denotes the concerned attacker, and Σ is a set of possible attack steps.

Countermeasures. There are different types of countermeasures to safeguard systems from potential attacks. These include system hardening mechanisms that raise the difficulty of launching successful attacks (e.g., timely patching), intrusion detection and prevention systems, as well as the resiliency mechanisms to limit the potential damage caused by invasions from attackers. We define countermeasures as a first-level element in CISSA to highlight their importance. In relation to the other elements in CISSA, a countermeasure is about what attacks it addresses and which parts of the system it hardens. When defining a concrete input scenario, we expect one to list the actual countermeasures that are in place.

4   Constructing Initial CISSA Cases

In this section we describe the six common input scenarios that we have defined using the proposed CISSA schema in Sect. 3. We challenge ourselves by selecting these scenarios from a diverse body of threats and application domains. We have put the specification files for all of our six initial scenarios online [1]. In Appendix A we illustrate the CISSA specification for one of the scenarios. We conclude this section by summarizing our experiences of constructing these initial CISSA cases.

Table 1. Characteristics of Example CISSA Scenarios

CISSA scenario	Unique characteristics
Stuxnet	Multi-step, several zero-day exploits, broken air-gap", command and control
Maroochy	Insider attack, poor auditing and access control policy, sabotage
Dragonfly	Targeted attack, watering hole, multi-step, trojanized software update, command and control
Target	Data breach, multi-step, integrated but insufficient security mechanisms, delayed incident response
SK communications	Highly targeted attack, data breach, poor security policy, trojanized software update
Syrian electronic army	Targeted attack, multi-step, evading detection and defense

4.1 A Diverse Set of CISSA Cases

We challenge ourselves to construct a diverse set of cases, with their characteristics summarized in Table 1. The following descriptions highlight for each of the six cases their unique aspects that our CISSA cases manage to incorporate.

Stuxnet. The Stuxnet attack [7,12], widely publicized in 2010, is arguably the most well-known example of a cyber attack targeting systems comprising of both cyber and physical equipment. The Stuxnet attack targeted the centrifuge machinery in Iranian nuclear enrichment facilities. The attack first enters into the cyber systems via public Internet and escalates its privilege, just as any ordinary cyber attack would do. This step involves multiple zero-day vulnerabilities. It then attempts to bypass the "air-gap" that is supposed to physically segregate the control networks from the Internet, by infecting USB storage devices. Once in the control network, the attacker intermittently changes the spinning speed of centrifuges, to cause lowered productivity and physical damage.

Maroochy. This scenario is based on the real-world incident that occurred in Maroochy Shire, Australia in 2000 [2]. In that attack, a disgruntled former employee stole a company laptop to send malicious radio commands to the sewage treatment system in Maroochy Shire. The attacker disguised his commands to appear as if coming from one of the pumping stations, causing pumps to malfunction; he also disabled alarms to conceal the attack. As result, sewage spilled at one of the pumping stations. The water treatment company noticed the "faults" and dispatched technicians to apply some countermeasures that were ineffective.

Dragonfly. The Dragonfly attack, publicized in 2014, was carried out by an advanced persistent threat actor [9]. The attack, which seems to target organizations from the energy and/or pharmaceutical sectors [13,28] involves the installation of Remote Access Tools, ostensibly for the purpose of information theft. Dragonfly was carried out via three attack vectors: (i) email spear phishing, (ii) a watering-hole attack designed to compromise industrial control system vendors' systems, and (iii) trojanized software designed to spread from the compromised vendors to their customers' systems.

Target. The data breach at Target Corporation in 2013 resulted in the loss of credit card data from 40 million customers [3,8]. This multi-stage attack began with the theft of an HVAC vendor's credentials for Target's vendor management web portal. From there, attackers were able to penetrate deep into Target's corporate network, steal personally identifiable information (PII) from a database, and deploy malware to pull customers' credit card information off of Target's point-of-sale machines. Stolen data were exfiltrated via FTP from inside the corporate network. We analyze this attack in greater detail as an example CISSA in Appendix A.

SK Communications. This scenario is based on a 2011 incident involving the theft of customer data from SK Communications: a South Korean internet service provider [6]. The attackers compromised a software vendor, and created a trojanized software update that installed a remote administration tool on more than 60 machines in the corporate network. The attackers leveraged this malware to exfiltrate personal information from over 35 million of SK Communications' customers.

Syrian Electronic Army. In 2013, a hacktivist group thought to have ties to the Syrian government initiated a number of phishing attacks against Western media organizations [16]. The attack follows two stages: (i) an initial phishing campaign from external email accounts to gain access to a user account in the target organization, and (ii) a second, internal, phishing campaign to obtain more desirable user accounts (e.g., with access to the company website's content management system). Once the attackers gain access to the company's website or social media account, they undertake web defacement or publication of material supporting their political agenda.

4.2   Our Experiences Constructing CISSA Cases

One lesson learned from the process of defining CISSA cases is the need for iteration. Since each of the real-world security incidents described above are unique, there was a need to progressively update the input specification. In fact, the *users* input U was added to the framework after identifying a gap in a previous ontology related to the modeling of human-centric attack vectors such as phishing. We believe future work and attention from the broader security community will further strengthen the CISSA framework and case descriptions.

A second critical issue is the level of detail to provide for each input class in the tuple. Real-world systems and cyber incidents are undoubtably complex, and model-based assessment tools often require different representations of the target system. We sought to address this by placing greater emphasis on *what*

Table 2. Model-based security assessment tools used in experiments

Tool	Category	Features
MulVAL [19]	Attack-graph-based	Integrating network and system vulnerability assessment
CySeMoL [27]	UML-based	Modeling various aspects of information system, with built-in knowledge base for quantitative evaluation
BDMP [22,23]	Attack-tree-alike, with extra modeling power	Modeling dynamic behaviors by Markov-chain-enhanced attack trees

Table 3. CISSA elements used as inputs to assessment tools

Tool	Network	Data	User	Operations	Undesirable Outcomes	Attack	Counter-measure
MulVAL	OVAL/Nessus/ Firewall scan reports	Nil	Built-in extensible library, e.g., hasAccount, etc.	Nil	Built-in extensible library, e.g., execCode, etc.	Nil	Nil
CySeMoL	Predefined extensible library, e.g., Network Zone, NetworkInterface, etc.	Predefined extensible library, e.g., Dataflow, Datastore, etc.	Predefined extensible library, e.g., Person, Account, etc.	Technical, operational, organizational levels modeled in templates	Nil	Predefined attack steps	Predefined defense steps
BDMP	User-constructed attack tree	User-constructed attack tree	Nil	Nil	Root of attack tree	Tree leaves, e.g., AA, ISE, TSE	Enable/ disable detection mode

should be specified rather than *how it should be represented*. While there may be no "right" level of detail, we believe one of the primary benefits of CISSA is making clear the starting point for various model-based security assessment efforts.

5   Putting CISSA into Use

As summarized earlier in Table 1, our small yet diverse set of scenarios exhibit some unique characteristics, e.g., on the nature and design of the system, the tactics, techniques, and procedures of the attack, and the response from the victim. Based on these characteristics, we group the six scenarios into three categories corresponding to different security features that are typically assessed: *technical, operational,* and *organizational*. As we discuss below, CISSA provides a convenient and concrete context for security analysts to explore technical, operational, and organizational security aspects, and to understand which security assessment tools or methods may be most appropriate for each one.

Such a common context enables comparison of different security assessment tools in their evaluations from various aspects. For instance, from the technical aspect, CISSA allows security analysts using assessment tools to answer questions such as *how do the network topology and configuration of the assessed system affect its security standing, for a particular scenario?* (**Experiment I**). With a given CISSA scenario with details in the systems, security analysts can tweak the network topology and configuration, and quickly re-evaluate the security of the resulted systems under the same settings. From the operational aspect, CISSA can assist analysts in answering questions like *how much would a better incident response procedure change the system's resilience against the given attack?* (**Experiment II**). For example, in the Target case, it took weeks before the victim confirmed the breach. If the response were more prompt, would the impact be lowered? Such hypothetical questions can be assessed by adopting different incident response options for the original CISSA case, and comparing the evaluation results from various assessment tools. In addition, CISSA can be useful in providing the context for evaluating questions like *how effective could a better security awareness program be at thwarting an attack in a given environment?* (**Experiment III**). Such analysis will be especially relevant to studying how cases such as Maroochy attack can be better prevented. Our experiments as detailed next focus on the three aspects, demonstrating how CISSA can be useful in comparing and informing different security assessment methodologies and tools.

5.1   Using CISSA to Compare and Inform Methodologies

Ultimately, the usefulness of CISSA to security practitioners and security assessment researchers depends on: (1) whether using CISSA provides insight on comparing and selecting existing security assessment tools; (2) whether using CISSA reveals aspects where existing assessment tools are doing well and/or lagging behind. To provide some initial insights into the usefulness of CISSA in the above aspects, we focus on three representative model-based security assessment methodologies with good tool support, among many others (e.g., ADTree, ADVISE, NETSPA). As summarized in Table 2, the three tools we choose, i.e., Multi-host, Multi-stage Vulnerability Analysis Language (*MulVAL*) [19], Cyber Security Modeling Language (*CySeMoL*) [27], and Boolean logic Driven Markov Process (*BDMP*) [22,23][1], use different formalisms to incorporate different security-related aspects, and subsequently (as we will show later), take in significantly different information and format as inputs. Table 3 describes how each tool takes in specific CISSA elements as input.

Feeding CISSA into the Tools. When conducting our experiments, we apply a *best-effort* approach, by (1) applying CISSA scenarios to assessment tools as much and as directly as possible, and (2) exploring the available features of the

---

[1] In this paper, we refer to the particular tool available at http://researchers.edf.com/software/kb3-44337.html.

tools as much as needed—as we will show, this often involves the use of some advanced features in the tools.

One practical benefit of having CISSA is to encourage different security assessment tools to unify their input formats. This would help security practitioners who need to work with different tools. Currently, our XML-based CISSA scenarios [1] cannot be directly used by the three tools in an automatic way. Instead, we manually convert the XML files into the appropriate input formats for each tool. From this exercise, we see that it should be feasible to automatically translate CISSA scenarios for use in some tools. For example, MulVAL has built-in parsers to collect input from certain vulnerability scanners, and CySeMoL supports XML import/export[2] However, it is less clear how to automate the procedure to convert CISSA scenarios for use in tools like BDMP, which is designed for manual use. One potential direction is to construct a basic attack tree as a starting point for BDMP by chaining together different attack steps using their pre-conditions and post-conditions.

While we strive to be comprehensive in CISSA, all three tools need extra information (beyond CISSA inputs) to conduct their assessment. For example:

– BDMP requires the probability distributions for launching an attack step. Since we cannot readily obtain such information from real-world measurements and reports, it is not included in CISSA.
– MulVAL computes the risk associated with different vulnerabilities by obtaining the corresponding vulnerability descriptions from a database.
– CySeMoL relies on built-in logical structures and probability relationships among different security-related concepts, in order to cater for non-expert users. While CySeMoL includes a meta-class editor to change these default settings, CISSA does not provide enough information to allow us to do so.

In the future, we believe that CISSA will evolve to provide more complete coverage for different tools. However, we foresee that there will always be certain tool-specific information that is not suitable for inclusion in CISSA. We believe that each security assessment should explicitly state the input information that was considered, and that any delta with the CISSA information should be made explicit.

In the remainder of this section, we use the above inputs (both inside and outside of CISSA) to conduct several tool-based experiments attempting to address three typical questions in the technical, operational, and organizational aspects. The purpose of these experiments is to show how CISSA can provide a realistic context for analyzing the processes used by the tools to gather inputs and produce these outcomes. It is not, however, intended to be a commentary on the soundness of the tools nor the accuracy of the outcomes.

---

[2] The PRM-based version in [27] provides XML import/export. However, we cannot find the format and semantics of the files, nor available tools for generating/processing them.

Experiment I (Technical Aspect). Here we challenge the tools to answer this question: how does the network topology and configuration of the assessed system affect its security standing for a particular scenario?

First, we use MulVAL to model the Stuxnet scenario, which is characterized by a sophisticated multi-step attack and the utilization of several vulnerability exploits as attack vectors. By design, MulVAL reports unpatched vulnerabilities in a network by leveraging various sources, such as OVAL, the Nessus scanner, NVD, etc. It integrates the discovery of such vulnerabilities into the assessment. To model the Stuxnet scenario, we find that we need to use some advanced features of MulVAL, e.g., creating new inference rules for generating attack graphs. For example, the default MulVAL engine cannot model USB-based penetration of the "air-gap". Hence, we add a few new rules, including:

```
interaction_rule(
 (execCode(Host, Perm) :-
  vulExists(Host,_,Software,_,privEscalation),
  localService(Host,Software,Perm),
  usbMounted(Host,USB_Drive),
  malwareLocated(USB_Drive)),
  rule_desc('Exploit via Infected USB',1.0)).
```

With these changes and the CISSA inputs, MulVAL generates a high-quality attack graph that closely resembles the attack graph used in an in-depth Stuxnet analysis report published by Tofino Security [4][3]. In the generated graph, attackers have more than 210 different possible attack paths to reach the process control network, by combinatorially exploiting seven different attack vectors over multiple network segments. MulVAL produces an attack success probability of 0.861 for this scenario.

To assess how network topology affects the system's security, we remove the vulnerable Web Navigator server in the topology to implement a "true air-gap", i.e., physical segregation between enterprise and control network, instead of firewall-isolated one. With this change, there are fewer than five attack paths still available to attackers, and the attack success probability produced by Mul-VAL decreases to 0.539 accordingly. We present the generated attack graphs for comparison in Figs. 3 and 4, respectively. We also experiment with two alternative configuration setups: one with all USB drives prohibited, and the other with remote employee access disabled. It turns out that security enhancement gained from these is less obvious than a true air-gap. MulVAL produces 0.825 and 0.667 attack success probabilities, respectively, for the two alternatives. This experiment shows how a tool like MulVAL can use information in CISSA to provide a concrete answer for the question posed earlier.

We also test BDMP and CySeMoL with the same Stuxnet scenario. For BDMP, unlike MulVAL which can automatically generate the attack graph, we have to manually construct the model. This is more time consuming, especially

---

[3] The main difference is that the hypothesis in [4] about Contractor Remote Access attack vector is not included in our CISSA case.

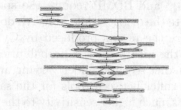

Fig. 3. Stuxnet: with Web Navigator server

Fig. 4. Stuxnet: no Web Navigator server. Figures 3 and 4 are presented to illustrate the different number of attack paths for the two settings, and not meant for showing the details.

Table 4. Results on attack success probability for Stuxnet scenario from CySeMoL and BDMP

Tool	10 days' attack	30 days' attack	180 days' attack
CySeMoL	.38	.43	.45
BDMP	.05	.34	.60

when we vary the network topologies and configuration setup of the system, such as removing the Web Navigator server to answer some "what if" questions. With CySeMoL, we also need to manually construct the network topologies for the assessed system. However, once it is constructed, it is fairly straightforward to alter it and re-run the assessment.

Differed Inputs, Differed Outputs: A key observation is that although these tools use similar metrics to represent the assessment result, i.e., the attack success probability, they produce vastly different values for the same system assessed. Part of the reason could be the presence of additional, tool-specific inputs, as discussed previously.

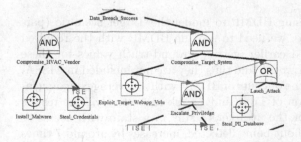

Fig. 5. Target data breach modeling with BDMP (partial)

Fig. 6. SEA modeling with CySeMoL (partial)

We now compare the attack success probabilities produced by the tools when we model the same Stuxnet scenario. Since both CySeMoL (the latest version of v2.3)[4] and BDMP require the specification of attack duration in order to compute the metric, while MulVAL does not have the notion of time, we can only compare the results produced by CySeMoL and BDMP meaningfully. Table 4 shows the attack success probabilities produced by CySeMoL and BDMP for the Stuxnet scenario under varying attack duration. As shown in the table, the results differ between tools for the same attack duration, and the BDMP tool exhibits much higher sensitivity to the attack duration. For security practitioners to better interpret differences like these, it is necessary to clearly specify the extra inputs used to conduct a security assessment.

Experiment II (Operational Aspect). Here we challenge the tools to answer this question: how much would a better incident response procedure change the system's resilience against the given attack?

We have varying degrees of success in conducting operational-level security assessment using the three tools. MulVAL does not have a built-in notion of time, nor about reactive defense procedures. Although we were able to introduce some customized rules for MulVAL in our last experiment, because the notions of time and reactive countermeasures are so fundamental, it is unclear how to enhance MulVAL to model these aspects. For CySeMoL, we can *indirectly* model the impact of reactive countermeasures by varying the attack duration parameter. We observe that the final risk value computed for the Target case is highly influenced by this parameter, e.g., the attack success probability can be reduced from 58 % to 14 % if the countermeasure can reduce the available time for the attacker from 1 week to 1 day.

In comparison, we find that some advanced features offered by BDMP [22] can be very useful in answering this question. In particular, BDMP allows a user to express the detection effectiveness for each individual attack step directly at four different stages: initial, ongoing, final, and a-posteriori. For the initial (when an attack step gets started) and final (when an attack step completes) steps, the user can specify a probability for the attack to be detected. For the ongoing (when an attack step is being carried out) and a-posteriori (when an attack step has been completed so the attacker can proceed to the next step if any), the user can specify the mean-time-to-detection assuming the time follows an exponential distribution.

We test these features by using BDMP to model the Target scenario (partially shown in Fig. 5). To do so, we need to provide BDMP with the required detection parameters. As argued earlier, getting ground-truth values for these parameters is difficult in the real world, and thus they are not included in CISSA. In our experiment, we compare the outputs of BDMP with an average ongoing/a-posteriori mean-time-to-detection of 1 day and 1 week respectively. The results show that the expected time for the attacker to stay in a state with access to the credit card information, without being detected, increases by around 7 times

---

[4] Earlier PRM-based versions of CySeMoL assume constant attack duration.

under the two settings. This modeling capability can be very useful for security practitioners to understand and promote the importance of rapid incident response, if the required estimation for time-to-detection can be obtained with reasonable accuracy.

While BDMP provides direct and detailed modeling for conducting these studies for the Target scenario, we find that as of now the tool can only deal with a one-time attack. This is insufficient to model repeated malicious attempts, such as in the Maroochy case. One possible enhancement to BDMP would be to integrate more complicated Markov models where a defeated attacker can adapt and re-launch the attack.

Experiment III (Organizational Aspect). Here we challenge the tools to answer this question: how effective could a better security awareness program be at thwarting the attack in the given environment?

We find that organizational aspects are not well modeled by the studied tools: only CySeMoL provides some simple modeling of a security awareness program. We thus created a model (partially shown in Fig. 6) in CySeMoL for the Syrian Electronic Army (SEA) scenario, exploring how the security awareness program would affect the security assessment. Specifically, for a model without the awareness program, CySeMoL produces a 28 % probability of success for the attacker to access the victim's Twitter account. After we enable the security awareness program for the targeted organization, the risk reduces to 20 %. On the other hand, a technical security mechanism like authentication protection can reduce the risk from the original 28 % to 17 %. CySeMoL allows the combined application of both organizational and technical security controls into the same model. For example, with both security controls applied, the risk is reduced to 7 %.

Although it is interesting to see the modeling of organizational security considerations together with other aspects in the same framework, we believe more organizational security details can be included: for example, the existence (or absence) of a policy for isolating different accounts (as in the SEA case), the effectiveness of the auditing procedures for employees who leave a company (as in the Maroochy case), or the trust policy for third-party vendors (as in the SK Communications case).

6 CISSA: Benefits, Limitations, and Roadmap

Our experiments using example scenarios demonstrate how CISSA provides a concrete context for analysis and identification of strengths and weaknesses of security assessment tools. The pros and cons we discussed and the quantitative values generated by the tools are provided only for the purposes of illustrating potential benefits of using CISSA. Although the results could be further refined, the process of using the different tools to answer specific security questions was illuminating, and we believe that CISSA can play an important role in the continued development of this research area.

6.1   Using CISSA to Advance the State-of-the-Art

From our evaluations we see three areas for improving model-based security assessment methodologies.

Clarifying Inputs. As we mention in Experiment I, differing inputs and input formats between different assessment tools has been a bottleneck, complicating meaningful comparison of tools or corroboration of results. Our sample CISSA scenarios have provided some initial but meaningful analysis on what different assessment tools seem to agree on, and where they differ in terms of inputs. When security practitioners need to conduct an assessment, they can have a better understanding about which tool or tools would be most suitable by looking at how they model similar scenarios.

Integrating Tools. Our experiments above demonstrate that different tools are designed to focus on specific aspects of security assessment. Enhancing each one individually to incorporate richer inputs is possible, albeit at the risk of making them more complicated and hence reducing their usability. A better alternative is to continue using different specialized tools to answer different questions, but to find certain ways to integrate them together, e.g., the way MulVAL integrates OVAL/Nessus scanning results. this front, common inputs are just the first step towards interoperability between tools. Common interfaces, consistent outputs, and methodologies or guidelines in partitioning assessment tasks also need to be established after inputs are agreed upon by different tools.

Modeling Security Beyond the Technical Level. All three tools we experiment with provide detailed modeling in the lower technical aspects. However, our success decreases as we move up to the operational and organizational level. Cross-reference between different levels is also preliminary in all three tools. By designing CISSA to contain information at all three aspects (and in a more balanced manner), CISSA can help promote a more holistic treatment of these different levels.

6.2   Present Limitations

The main limitation of our current CISSA definition is that it does not guarantee on the completeness of information. However, the idea is to populate the CISSA repository with such information as commonly required by popular security assessment tools. CISSA can be readily extended to include additional information about systems being assessed or the details about the threats as necessary. However, CISSA at least makes input to security assessment explicit, which can significantly avoid possible misunderstandings and misassumptions on how systems are assessed.

Another potential limitation is that the conversion from CISSA cases to tool-specific inputs is non-trivial. As a one-time effort, once a particular CISSA case has been converted, it can be readily reused for similar scenarios by minor edits. We plan to develop tools to help automate such processes. Our hope is the usefulness of CISSA demonstrated in practice will motivate gradual and eventual converge in the format of inputs to different security assessment tools.

A third limitation could lie in the process of security analysts selecting most relevant CISSA cases for the systems under assessment, when the CISSA repository grows much richer. In fact, we would encourage a worst-case analysis by considering all attack scenarios. Ruling out security vulnerabilities is by itself very challenging.

6.3   The Future of CISSA

Pinpointing areas where model-based security assessment can improve is one of the key roles we believe CISSA can play. At the same time, in better shaping security assessment methodologies we expect CISSA itself to evolve.

Our initial experiences in constructing sample CISSA scenarios indicate that CISSA is more of an evolving effort than a static collection. The seven-element CISSA framework we propose is also to be iteratively refined with additional effort on constructing additional sample CISSA scenarios, e.g., in terms of whether to include/exclude some information, how to define the delta accurately, and how to enable automatic translation. We understand this has to be a community-wide effort, and we plan to incorporate CISSA modeling and CISSA enabled comparison with other tools in the development of our own security assessment software [30]. Furthermore, we plan to organize workshops to provide the forum for developers of security assessment tools to discuss how to make CISSA more useful. We hope other security researchers will share our vision, so we can work together to continue building an ever-richer CISSA repository.

We believe this type of sustained community effort will be the first step in making model-based security assessment more scientific and valuable. Establishing a commonly accepted input repository for security assessment will form a basis for benchmarking security assessment tools. We have seen benchmarks in many other areas, such as the JavaScript engine, GPU rendering, routing protocols, etc. Unfortunately, we are not yet close to having common benchmarks to gauge model-based security assessment tools, e.g., for evaluating their breadth in taking all relevant information into account, their sensitivity to network topological changes, their run-time performance to assess a particular scenario, etc. Security researchers first need to largely agree on the inputs, before they can work out how the outputs from security assessment tools can be interpreted as benchmarking results.

7   Conclusion

This paper presents an open repository of common input scenarios for security assessment (CISSA). By constructing six sample input scenarios and experimenting with three assessment tools, we show the potential usefulness of CISSA for security practitioners and for the future development of security assessment tools. In particular, well specified common inputs could facilitate the comparison and integration of different assessment tools. We hope our work will promote a concerted community effort to build a richer repository of CISSA for supporting model-based security assessment studies.

Acknowledgements. This study is supported by the research grant for the Human-Centered Cyber-physical Systems Programme at the Advanced Digital Sciences Center from Singapore's Agency for Science, Technology and Research (A*STAR).

A Appendix: CISSA Example

In this section we provide an example CISSA by describing the seven elements (N, D, U, O, X, A, C) for the Target Corporation data breach. The input files themselves are available online in XML format [1].

A.1 Brief Recap of the Target Incident

The data breach at Target Corporation in 2013 resulted in the loss of credit card data from 40 million customers [3,8]. This multi-stage attack began with the theft of an HVAC vendor's credentials for Target's vendor management web portal. From there, attackers were able to penetrate deep into Target's corporate network, steal personally identifiable information (PII) from a database, and deploy malware to pull customers' credit card information off of Target's point-of-sale machines. Stolen data were exfiltrated via FTP from inside the corporate network.

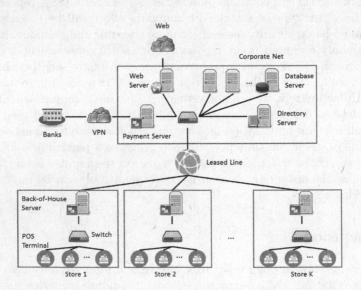

Fig. 7. Components and network for target input scenario

A.2 CISSA Definition for the Target Scenario

To illustrate how CISSA specification can represent the details of read-world incidents for security assessment, we briefly present how we define the Target scenario under CISSA.

Components and Network N. A large retailer such as Target typically has 100s or 1000s of locations, each with numerous point-of-sale (POS) stations that accept and process customer payments. Figure 7 depicts the network connections (E_N) and devices (V_N) considered in this scenario. The model includes K store locations, each with T_K POS machines which are connected via a switch to a back-of-house (BoH) server. This BoH server connects to a central payment server at the corporate network, which interfaces with external financial institutions to verify transactions. The corporate network also includes a directory server, a web server, and a database server. We abstract other corporate services from this model. In the XML files online [1] we specify additional information.

Data D. Data plays a central role in the Target CISSA. The attackers' goal was the theft of customer data, and this was made possible by the acquisition of additional system-specific data. Table 5 describes the data items that are modeled in this scenario. We provide a unique identifier for each data item ID_D, a description of the data's properties L_D, and a list of devices that interact with the data Map_D (network links are omitted for brevity).

Table 5. Data items in the target input scenario

Identifier (ID_D)	Description (L_D)	Mapping (Map_D)
D1	Credit card number	POS terminal, BoH server, Payment server, Bank
D2	Customer PII (e.g., name, address)	POS terminal, BoH server, Payment server, Bank, Database Server
D3	Admin access token	Servers in corporate network
D4	Active Directory listing	Directory server

Users U. In this scenario, several different user types are affected by the incident. In particular, credentials from a *vendor* and, later, a *Domain Administrator*, enabled the adversaries to steal sensitive information relating to the company's *customers*. Table 6 summarizes the relevant user information.

Operations O. Arguably the most important system operation in this scenario is the handling of consumer credit card information during POS transactions. Figure 8 depicts this process. Here the vertices V_O denote the major operations from the POS terminal, a store's BoH Server, and the Bank responsible for clearing the transaction, while the edges E_O imply sequential order. The mapping function Map_O in this case assigns specific devices to the roles described above

Table 6. Users in the target input scenario

Identifier (ID_U)	Description (L_U)	Access (Map_U)
$U1$	Contractor with vendor web portal account	Web server
$U2$	Domain Administrator for corporate network	All devices/links in N
$U3$	Customer in a store	POS terminal

(e.g., POS Terminal 5 in Store #300). Additional system operations in this scenario could include the POS Terminal or BoH Server's software update process, or the processes for collecting and storing personally identifiable information (PII) in the company's database.

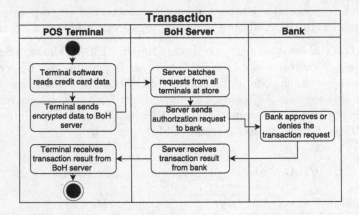

Fig. 8. Transaction operations for target input scenario

Undesirable Outcomes X. In this scenario we model the final undesirable outcomes of the attack as *loss of credit card data* and *loss of personally identifiable information*, as specified in Table 7. Due to space limit, we do not elaborate on intermediate undesirable outcomes x_1 to x_7 for each attack step, while providing brief summary in Table 8.

Attack A. The attacker input α is modeled with:

- **Goal:** theft of credit card data $X1$.
- **Access:** external attacker, with access to "Web" in N.
- **Skills:** use of existing tools and malware (no zero-days).

Table 7. Undesirable outcomes in the target scenario

Identifier (ID_X)	Description (L_X)	Mapping (Map_X)	Implications (Imp_X)
$X1$	Loss of credit card data	POS machine	Fraud, legal liability
$X2$	Loss of PII	Database server	Legal liability, loss of goodwill

Table 8. Attack steps for the target input scenario

Attack step (L_σ)	Pre-condition (Pre_σ)	Post-condition ($Post_\sigma$)
1. Steal credentials	<(Vendor's network access), (Server vulnerability exploiting techniques)>	<(Credentials of Target's systems), (), x_1 (Credential leak)>
2. Expl. web server	<(Credentials of Target's systems), (Server vulnerability exploiting techniques)>	<(Privilege to execute OS commands), (), x_2 (Privilege leak)>
3. Steal token	<(Access to Target's servers), (Know-how of collecting NT hashes from memory)>	<(Corporate network admin privilege), (), x_3 (Privilege escalation)>
4. Create account	<(Admin privilege to add new user to Domain), ()>	<(Access to corporate network), (), x_4 (Malicious admin account)>
5. Steal PII	<(Access to corporate network), (Skill to use database server)>	<(Access to customer records), (), x_5 (Unauthorized access)>
6. Install malware	<(Access to POS machines' writable folders), (Malware infection capabilities)>,	<(Access to data on POS), (), x_6 (Malware infection)>
7. Aggregate data	<(Access to FTP servers in corporate network, access to sensitive data), (Basic file transfer techniques)>	<(), (), x_7 (Sensitive data aggregated>
8. Exfiltrate data	<(Access to outward-facing internet connection, access to sensitive data), (Skills to stealthily exfiltrate files)>	<(), (), $X1 \cup X2$ (Data leak)>

The attack on Target's corporate network and POS system is thought to consist of 11 steps [3]. We model a simplified 8-step attack, i.e., σ_1: Steal credentials of vendor, σ_2: Exploit vulnerability on Target web portal, σ_3: Steal Domain Admin access token, σ_4: Create new Domain Admin account, σ_5: Steal PII from database, σ_6: Install malware on POS machines, σ_7: Aggregate stolen data in network, and σ_8: Exfiltrate data via FTP.

The above text constitutes the attack step description input (L_ω). In Table 8 we specify the pre-conditions (Pre_σ) and post-conditions ($Post_\sigma$) for these attack steps.

Countermeasures C. The credit card industry maintains a set of standards for data protection [20]. In addition to those guidelines—which were followed in this scenario—other countermeasures can potentially detect or prevent similar attacks.

Table 9. Countermeasures in the target input scenario

Identifier (ID_C)	Description (L_C)	Mapping (Map_C)
$C1$	Multi-factor authentication	Users $U1$, $U2$
$C2$	Application whitelisting	POS terminal
$C3$	Real-time monitoring	Directory server, Web server

- **Multi-factor authentication** for the outward-facing vendor portal, and for the Domain Administrators.
- **Application whitelisting** for the point-of-sale machines and the servers involved in transaction verification.
- **Real-time monitoring** of user lists and network queries to detect the addition of new user accounts (particularly admin accounts) and potentially identify lateral movement of an attacker within the network.

References

1. Public Repository for CISSA. http://www.illinois.adsc.com.sg/cissa
2. Abrams, M., Weiss, J.: Malicious control system cyber security attack case study - Maroochy water services, Australia (2008)
3. Aorato Labs: The untold story of the target attack step by step, August 2014. http://www.aorato.com/blog/untold-story-target-attack-step-step/
4. Byres, E., Ginter, A., Langill, J.: How stuxnet spreads - a study of infection paths in best practice systems. www.tofinosecurity.com/how-stuxnet-spreads
5. Chen, B., Kalbarczyk, Z., Nicol, D.M., Sanders, W.H., Tan, R., Temple, W.G., Tippenhauer, N.O., An Hoa, V., Yau, David, K.Y.: Go with the flow: toward workflow-oriented security assessment. In: NSPW (2013)
6. Command Five Pty Ltd.: SK Hack by an Advanced Persistent Threat, September 2011. http://www.commandfive.com/papers/C5_APT_SKHack.pdf
7. Falliere, N., Murchu, L.O., Chien, E.: Symantec security response: W32.stuxnet dossier. www.symantec.com/content/en/us/enterprise/media/security_response/whitepapers/w32_stuxnet_dossier.pdf
8. iSightPartners: Kaptoxa point of sale compromise, January 2014. http://www.securitycurrent.com/resources/files/KAPTOXA-Point-of-Sale-Compromise.pdf
9. Kaspersky Lab Global Research and Analysis Team: Energetic bear - crouching yeti, July 2014. http://securelist.com/files/2014/07/EB-YetiJuly2014-Public.pdf
10. Kordy, B., Mauw, S., Radomirović, S., Schweitzer, P.: Foundations of attack-defense trees. In: FAST, pp. 80–95 (2011)
11. Kordy, B., Pietre-Cambacedes, L., Schweitzer, P.: DAG-based attack, defense modeling: don't miss the forest for the attack trees (2013). CoRR arXiv:1303.7397
12. Kriaa, S., Bouissou, M., Pietre-Cambacedes, L.: Modeling the stuxnet attack with BDMP: towards more formal risk assessments. In: Proceedings of International Conference on Risk and Security of Internet and Systems (CRiSIS), pp. 1–8, October 2012
13. Langill, J.: Defending against the dragonfly cyber security attacks (2014). http://www.belden.com/docs/upload/Belden-White-Paper-Dragonfly-Cyber-Security-Attacks.pdf

14. LeMay, E., Ford, M., Keefe, K., Sanders, W.H., Muehrke, C.: Model-based security metrics using ADversary VIew Security Evaluation (ADVISE). In: QEST (2011)
15. Lippmann, R.P., Ingols, K.W.: An annotated review of past papers on attack graphs (2005)
16. Mandiant, a FireEye Company: Beyond the breach (2014). https://dl.mandiant.com/EE/library/WP_M-Trends2014_140409.pdf
17. North American Electric Reliability Corporation: Critical infrastructure protection standards. http://www.nerc.com/pa/Stand/Pages/CIPStandards.aspx
18. Ou, X., Boyer, W.F.: A scalable approach to attack graph generation. In: CCS (2006)
19. Xinming, O., Govindavajhala, S., Appel, A.W.: Mulval: a logic-based network security analyzer. In: USENIX Security (2005)
20. PCI Security Standards Council: PCI SCC data security standards overview. https://www.pcisecuritystandards.org/security_standards/
21. Phillips, C., Swiler, L.: A graph-based system for network-vulnerability analysis. In: NSPW (1998)
22. Piètre-Cambacédès, L., Bouissou, M.: Attack and defense modeling with BDMP. In: Kotenko, I., Skormin, V. (eds.) MMM-ACNS 2010. LNCS, vol. 6258, pp. 86–101. Springer, Heidelberg (2010)
23. Pietre-Cambacedes, L., Bouissou, M.: Beyond attack trees: dynamic security modeling with Boolean logic driven Markov processes (BDMP). In: EDCC (2010)
24. Sanders, W.: Quantitative security metrics: unattainable holy grail or a vital breakthrough within our reach? IEEE-SPM **12**, 67–69 (2014)
25. Schneier, B.: Attack trees: modeling security threats. Dr. Dobb's J. **24**, 21–29 (1999)
26. Sheyner, O., Haines, J., Jha, S., Lippmann, R., Wing, J.: Automated generation and analysis of attack graphs. In: IEEE S&P (2002)
27. Sommestad, T., Ekstedt, M., Holm, H.: The cyber security modeling language: a tool for assessing the vulnerability of enterprise system architectures. IEEE Syst. J. **7**(3), 363–373 (2013)
28. Symantec Security Response: Dragonfly: cyberespionage attacks against energy suppliers, July 2014. http://www.symantec.com/content/en/us/enterprise/media/security_response/whitepapers/Dragonfly_Threat_Against_Western_Energy_Suppliers.pdf
29. Verendel, V.: Quantified security is a weak hypothesis. In: NSPW (2009)
30. Vu, A.H., Tippenhauer, N.O., Chen, B., Nicol, D.M., Kalbarczyk, Z.: CyberSAGE: a tool for automatic security assessment of cyber-physical systems. In: Norman, G., Sanders, W. (eds.) QEST 2014. LNCS, vol. 8657, pp. 384–387. Springer, Heidelberg (2014)

Differential Privacy Analysis of Data Processing Workflows

Marlon Dumas[1], Luciano García-Bañuelos[1(✉)], and Peeter Laud[2]

[1] University of Tartu, Tartu, Estonia
{marlon.dumas,luciano.garcia}@ut.ee
[2] Cybernetica, Tallinn, Estonia
peeter.laud@cyber.ee

Abstract. Differential privacy is an established paradigm to measure and control private information leakages occurring as a result of disclosures of derivatives of sensitive data sources. The bulk of differential privacy research has focused on designing mechanisms to ensure that the output of a program or query is ϵ-differentially private with respect to its input. In an enterprise environment however, data processing generally occurs in the context of business processes consisting of chains of tasks performed by multiple IT system components, which disclose outputs to multiple parties along the way. Ensuring privacy in this setting requires us to reason in terms of series of disclosures of intermediate and final outputs, derived from multiple data sources. This paper proposes a method to quantify the amount of private information leakage from each sensitive data source vis-a-vis of each party involved in a business process. The method relies on generalized composition rules for sensitivity and differential privacy, which are applicable to chained compositions of tasks, where each task may have multiple inputs and outputs of different types, and such that a differentially private output of a task may be taken as input by other tasks.

1 Introduction

The broad availability of rich consumer data is driving businesses to become increasingly data-driven in their daily operations. In particular, it is becoming common practice for businesses to exploit private data about their current or potential customers to design, sell and deliver services. As a broader set of organizational stakeholders become involved in processing personal customer data – sometimes across organizational boundaries – it becomes increasingly critical to measure and control private information leakages.

Differential privacy [5] has emerged as a promising foundation to quantify and control private information leakages stemming from access to sensitive data sources. The bulk of research in this field has focused on designing mechanisms to ensure that the output of a given program or query is $\epsilon-$differentially private with respect to a collection of input objects, for a given privacy budget ϵ. In other words, the contribution of each object in the input collection to the output is bounded by a term dependent on the privacy budget.

B. Kordy et al. (Eds.): GraMSec 2016, LNCS 9987, pp. 62–79, 2016.
DOI: 10.1007/978-3-319-46263-9_4

In an enterprise environment however, data processing generally occurs in the context of business processes consisting of complex chains of tasks performed by a range of IT system components and human actors. Ensuring privacy in this setting requires us to reason not only in terms of an individual disclosure of the output of a program or query to one party, but rather in terms of series of disclosures to a range of parties.

This paper addresses the problem of analyzing differential privacy in the context of business processes that involve multiple tasks, such that the output of one task may be used as input by other tasks, and such that intermediate or final outputs are disclosed to multiple parties. The paper addresses this problem in the setting where business processes are codified using graphical models consisting of processing nodes that extract data from potentially sensitive data sources, transform the extracted data, and disclose derivatives thereof using a differential privacy mechanism. Given such a graphical model, the paper outlines a technique to address the following question: *How much information about individual objects in each input data collection does one execution of a process model disclose to each involved party?*

To illustrate this problem, we consider a simplified process to produce a report about combined (data and call) service usage at a telecommunication services provider. This process is depicted in Fig. 1 using the standard Business Process Model and Notation (BPMN) [13]. The telco provider is represented by a pool.[1] There is a separate pool below it, corresponding to a contractor hired by the telco to provide business consultancy services. To provide its services, the contractor needs to access weekly "service summary reports" produced by the telco. Inside the telco's pool, there are two roles represented by the lanes labeled "Data Analyst 1" and "Data Analyst 2". The process starts when a new summary report is created (cf, the start event labelled "summary report required"). First, Data Analyst 1 performs a task wherein a set of call records are accessed in order to prepare a call summary table. We assume that this collection of call records (represented by a data collection – rectangle with a folded corner and three vertical stripes) contains sensitive data. Hence, Data Analyst 1 does not get the actual data collection, but only the result of a differentially private query. As a result of this task, a "Call summary table" is produced. Next, an automated task is executed that combines the previous "Call summary table" with another collection of "Data connection records" in order to produce a "Combined report". Again, since collection "Data connection records" contains sensitive data records, the program executing this latter task incorporates a differential privacy mechanism, which ensures that the combined report is ϵ_2- (resp. ϵ_3-) differentially private with respect to "Data connection records" (resp. "Call summary table"). The combined report is then checked by Data Analyst 2, who may modify it. The process ends with a "message event" denoting the fact that the combined report is sent out to the contractor.

---

[1] A pool in BPMN (represented by a horizontal rectangle) represents an independent organizational entity that communicates with other entities via *message flows*, represented via dashed arrows.

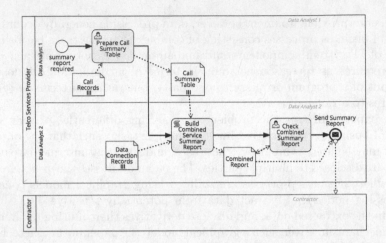

Fig. 1. Model of a report preparation process (in BPMN).

Given a graphical process model annotated with metadata about differentially private data releases, the technique proposed in this paper calculates for each stakeholder in the process (Data Analyst 1, Data Analyst 2 and Contractor) and for each input of the process (here the "Call records" and the "Data connection records"), how much ϵ privacy budget the stakeholder consumes with respect to the data collection in question during one execution of the process. This output can be used by an analyst to fine-tune the process (e.g. by adjusting the ϵ_i privacy budgets) in order to achieve a certain level of privacy vis-a-vis of each stakeholder and each data collection.

The proposed technique relies on a theoretical foundation that provides composition rules to calculate sensitivity and differential privacy of chained compositions of tasks, where these tasks take multiple inputs and produce multiple outputs of different types. The paper outlines an algorithm that applies these composition rules iteratively in order to calculate end-to-end differential privacy for a given process.

The rest of the paper is structured as follows. Section 2 introduces concepts and associated notation used subsequently. Section 3 presents the definitions of differential privacy and sensitivity and associated composition rules. Next, Sect. 4 introduces a notation for privacy-enhanced process modeling and presents an algorithm for differential privacy analysis of such models. Finally Sect. 5 discusses related work and Sect. 6 summarizes the contribution and outlines directions for future work.

2   Notation and Preliminaries

We use $\mathbb{R}$ and $\mathbb{N}$ to denote the sets of real and natural numbers, respectively. The sets of non-negative real and extended real numbers are denoted by $\mathbb{R}_+$ and $\mathbb{R}_+^\infty = \mathbb{R}_+ \cup \{\infty\}$. If $a, b \in \mathbb{R}$, then $[a, b]$ denotes the set $\{x \in \mathbb{R} \,|\, a \leq x \leq b\}$.

If X is a set then $\mathcal{D}(X)$ denotes the set of all *countable* probability distributions of X. The elements of $\mathcal{D}(X)$ are mappings $\chi : X \to [0, 1]$, such that the set $\mathrm{supp}(\chi) = \{x \in X \mid \chi(x) > 0\}$ is (at most) countable.

Given a probability distribution $\psi \in \mathcal{D}(X \times Y)$, we let $\mathsf{proj}_1 \psi \in \mathcal{D}(X)$ and $\mathsf{proj}_2 \psi \in \mathcal{D}(Y)$ denote its projections to the first and second component, respectively. These are defined by $\mathsf{proj}_1 \psi(x) = \sum_{y \in Y} \psi(x, y)$ for all $x \in X$ and similarly for $\mathsf{proj}_2 \psi$, where the sum is well-defined due to the support of ψ being countable. For $\chi \in \mathcal{D}(X)$ and $\phi \in \mathcal{D}(Y)$ we let $\chi \otimes \phi \subseteq \mathcal{D}(X \times Y)$ denote the set of all such probability distributions ψ that satisfy $\mathsf{proj}_1 \psi = \chi$ and $\mathsf{proj}_2 \psi = \phi$.

Let $f : X \to \mathcal{D}(Y)$ and $g : Y \to \mathcal{D}(Z)$. There is an obvious way to "compose" f and g, the result of which we denote with $g \circ_{\mathsf{KI}} f$ and which is defined by

$$\Pr[(g \circ_{\mathsf{KI}} f)(x) = z] = \sum_{y \in Y} \Pr[f(x) = y] \cdot \Pr[g(y) = z] \tag{1}$$

for all $x \in X$ and $z \in Z$.

The notion of sensitivity of mappings used in this paper relies on (extended) metric spaces defined as follows.

Definition 1 (Metric Space). *A metric space is a set X together with a metric d_X on it. A mapping $d_X : X \times X \to \mathbb{R}_+$ is a metric if it satisfies the following conditions:*

- *for all $x, y \in X$: $d_X(x, y) = 0$ iff $x = y$;*
- *for all $x, y \in X$, $d_X(x, y) = d_X(y, x)$;*
- *for all $x, y, z \in Z$, $d_X(x, z) \leq d_X(x, y) + d_X(y, z)$.*

An *extended metric* may also take the value ∞. An *extended metric space* is a set X together with an extended metric on it.

3 Differential Privacy

Let $\mathcal{R}$ be the set of possible database records and $X = \mathbb{N}^{\mathcal{R}}$ be the set of databases (i.e. a database is a multiset of records). Let $\mathcal{O}$ be a set of possible outcomes and $\mathcal{M} : X \to \mathcal{O}$ a probabilistic map (an *information release mechanism*). For $r \in \mathcal{R}$ let $x_1 \overset{r}{\sim} x_2$ denote that x_1, x_2 *differ only by r*, i.e. $x_1(r) = x_2(r) \pm 1$ and $x_1(r') = x_2(r')$ for all $r' \in \mathcal{R} \backslash \{r\}$. Two databases $x_1, x_2 \in X$ are *adjacent* if $x_1 \overset{r}{\sim} x_2$ for some $r \in \mathcal{R}$. Let d_X be any (extended) metric on X.

Definition 2 (Differential Privacy [5]). *Let $\varepsilon \in \mathbb{R}$. The mechanism $\mathcal{M}$ is ε-differentially private if $\Pr[\mathcal{M}(x_1) \in S] \leq e^{\varepsilon} \cdot \Pr[\mathcal{M}(x_2) \in S]$ for all $S \subseteq \mathcal{O}$ and all adjacent databases $x_1, x_2 \in \mathcal{X}$.*

There are a number of ways to make information release mechanisms private, but the most commonly used techniques amount to adding a certain amount of noise to the output of the mechanism. The noise has to be sampled from the correct distribution, in order to obtain the bounds on the ratio of probabilities, as demanded by Definition 2. The Laplacian distribution has the necessary properties [6]. The required magnitude of the noise depends on the function that is computed by the mechanism. A function that may have very different outputs for databases differing only a little requires more noise to be added than a function that changes only slowly.

Sensitivity is a key tool to reason about the differential privacy of information release mechanisms. It gives upper bounds for the ratio of the change in the value of the function with respect to a change in the argument of the function. For mechanisms that first compute a "useful" function and then add noise to it, the differential privacy of the resulting mechanism is the ratio of the sensitivity of that function and the magnitude of the added noise.

Definition 3 (Sensitivity). *Let X and Y be two metric spaces with distances d_X and d_Y on them. Let $c \in \mathbb{R}_+$. We say that a function $f : X \to Y$ is c-sensitive, if for all $x_1, x_2 \in X$, the inequality $d_Y(f(x_1), f(x_2)) \leq c \cdot d_X(x_1, x_2)$ holds.*

Differential privacy itself can also be seen as an instance of sensitivity. Indeed, define the following extended metric d_{dp} over $\mathcal{D}(Y)$:

$$d_{\mathrm{dp}}(\chi, \chi') = \sup_{y \in Y} \left| \ln(\chi(y) / \chi'(y)) \right|.$$

Then a mechanism $\mathcal{M}$ from X to Y is d_X-private iff it is 1-sensitive with respect to the distances d_X on X and d_{dp} on $\mathcal{D}(Y)$.

The well-known composition theorems of differential privacy are instantiations of more general results on sensitivity of composed mappings. We start with the simplest result for sensitivity.

Proposition 1. *Let $f : X \to Y$ be c-sensitive with respect to the distances d_X on X and d_Y on Y. Let $f' : Y \to Z$ be c'-sensitive with respect to the distances d_Y on Y and d_Z on Z. Then $f' \circ f : X \to Z$ is $c \cdot c'$-sensitive with respect to the distances d_X on X and d_Z on Z.*

Proof. Let $x, x' \in X$. Then $d_Z(f'(f(x)), f'(f(x'))) \leq c' \cdot d_Y(f(x), f(x')) \leq c' \cdot c \cdot d_X(x, x')$.

This proposition can be generalized to multivariate mappings. Let $i \in \{1, \ldots, n\}$. We say that a mapping $f' : Y_1 \times \cdots \times Y_n \to Z$ is $c'_i - sensitive in its i - th argument$, if for all tuples $(y_1, \ldots, y_{i-1}, y_{i+1}, \ldots, y_n) \in Y_1 \times \cdots \times Y_{i-1} \times Y_{i+1} \times \cdots \times Y_n$, the univariate mapping $f(y_1, \ldots, y_{i-1}, \cdot, y_{i+1}, \ldots, y_n)$ is c'_i-sensitive.

Proposition 2. *For $i \in \{1, \ldots, n\}$, let $f_i : X \to Y_i$ be c_i-sensitive with respect to the distances d_X on X and d_{Y_i} on Y_i. Let $f' : Y_1 \times \cdots \times Y_n \to Z$ be c'_i-sensitive with respect to the distances d_{Y_i} on Y_i and d_Z on Z (for all $i \in \{1, \ldots, n\}$). Then the mapping $g : X \to Z$, defined by $g(x) = f'(f_1(x), \ldots, f_n(x))$, is $\sum_{i=1}^{n} c_i c'_i$-sensitive with respect to the distances d_X on X and d_Z on Z.*

Proof. Let $x, x' \in X$. Let $z_i = f'(f_1(x), \ldots, f_i(x), f_{i+1}(x'), \ldots, f_n(x'))$. Then $z_0 = g(x')$, $z_n = g(x)$ and by Proposition 1, $d_Z(z_{i-1}, z_i) \leq c_i c_i' \cdot d_X(x, x')$. The claim of the proposition follows from the triangle inequality.

The sequential composition theorem for differential privacy [12, Theorem 3] is really just a special case of Proposition 2. In their setting, there is a dataset $x \in X$ and information release mechanisms $\mathcal{M}_1$ and $\mathcal{M}_2$, which are respectively ε_1- and ε_2-differentially private. Let the possible set of outcomes of $\mathcal{M}_i$ be M_i. First $\mathcal{M}_1$ and then $\mathcal{M}_2$ are invoked on x; the exact invocation of $\mathcal{M}_2$ may depend on the result of $\mathcal{M}_1$. Finally, the result of $\mathcal{M}_2$ is published. This result may include the result of $\mathcal{M}_1$, because it affected the invocation of $\mathcal{M}_2$. Such composition of $\mathcal{M}_1$ and $\mathcal{M}_2$ is shown to be $(\varepsilon_1 + \varepsilon_2)$-differentially private.

Propostion 2 applies to this setting in the following manner. We have $\mathcal{M}_1 : X \to \mathcal{D}(M_1)$ and $\mathcal{M}_2 : X \times M_1 \to \mathcal{D}(M_2)$. Let $\overline{\mathcal{M}_2} : X \times \mathcal{D}(M_1) \to \mathcal{D}(M_2)$ be the lifting of $\mathcal{M}_2$ to probability distributions in its second argument:

$$\Pr[\overline{\mathcal{M}_2}(x, \chi) = m_2] = \sum_{m_1 \in M_1} \chi(m_1) \Pr[\mathcal{M}_2(x, m_1) = m_2].$$

Consider the Hamming distance on X (two datasets in X are *adjacent* iff their distance is 1), and the distance d_{dp} on both $\mathcal{D}(M_1)$ and $\mathcal{D}(M_2)$. Let $f_1 = id_X$, $f_2 = \mathcal{M}_1$ and $f' = \overline{\mathcal{M}_2}$. The sensitivity of f_1 is 1, the sensitivity of f_2 is ε_1, and the sensitivities of f' in its first and second argument are ε_2 and 1, respectively. The latter follows the fact that no post-processing of a differentially private query can lower the privacy guarantees it provides. We now apply Proposition 2 and find that the composition of f' with $f_1 \times f_2$ is $(\varepsilon_1 + \varepsilon_2)$-sensitive. This composition corresponds to the invocation of $\mathcal{M}_1$ and $\mathcal{M}_2$ one after another, as described above.

4   Privacy Analysis of Data Processing Workflows

In this section, we introduce a graphical notation for capturing data processing workflows with differential privacy, and we define algorithms to analyze the end-to-end differential privacy of such workflows.

4.1   Data Processing Workflows

In Sect. 1, we presented a motivating example of a business process using the BPMN notation. While BPMN is a widely used standard, it is also rather complex. It comprises several dozen types of notational elements, covering several flavors of parallel and conditional branching, sequential and parallel repetition, exception handling and transactional constructs among others. For privacy analysis purposes, we propose to reason on a simpler and more abstract graphical notation, herein called *data processing workflow*. This simpler process modeling notation is focused on capturing how data sources taken as input by a business process are transformed into intermediate and final outputs, each of which is disclosed to one or multiple parties. Below we introduce data processing workflows without considering the notion of "disclosure to a party". The latter notion is added in Sect. 4.3.

A data processing workflow consists of *data nodes, processing nodes* and *data-flow arcs*. A data-flow arc connects a data node to a processing node or vice-versa. A data node without any incoming arc is called a *source data node*. It corresponds to an object or collection of objects that are given as input to the workflow. A data node without any outgoing arc is called an *output node* (i.e. it is data produced by an execution of the workflow). A data node with both incoming and outgoing arcs is called an *intermediate node*.

Figure 2 shows an example of a data processing workflow. Processing nodes are represented as rounded rectangles, while data nodes are rectangles with their top-right corner folded over.

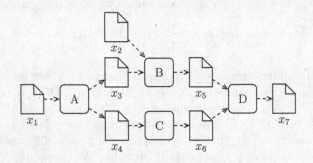

Fig. 2. Example of a data processing workflow

Formally, a *Data Processing Workflow* W is a tuple (D, P, F), where D and P are two finite, disjoint sets, and F is a relation on $(D \times P) \cup (P \times D)$. For convenience, we will refer to $D \cup P$ as the set of nodes N. The elements of D are data nodes and the elements of P are processing nodes, that is, nodes representing computations over some input data.

Given a node $n \in N$, we define $\bullet n = \{m \mid (m, n) \in F\}$ (the predecessors of n) and $n\bullet = \{m \mid (m, n) \in F\}$ (the successors of n). A workflow W is said *well-formed* if it induces an acyclic, weakly connected graph, with the following additional restrictions: every node $d \in D$ has at most one successor and at most one predecessor, i.e. $\mid \bullet d \mid \leq 1$ and $\mid d\bullet \mid \leq 1^2$, and every node $p \in P$ has at least one predecessor and at least one successor node, i.e. $\mid \bullet p \mid \geq 1$ and $\mid p\bullet \mid \geq 1$. In the following, we consider only well formed workflows.

---

[2] Acyclicity is also required on data dependencies as way to simplify the presentation. However, this restriction does not affect the generality of our approach. A cyclic data dependency would usually stand for a data update access. The same intuition can alternatively be represented with two data nodes: one data node representing the read data object and the other one representing the written data.

A privacy-enhanced workflow is a workflow annotated with differential privacy and sensitivity values, which we assume are derived separately via an analysis of a program or query implementing the data processing node (as discussed later in Sect. 5). Formally, a *Privacy-enhanced workflow* is a tuple $(W, \mathcal{E}, \mathcal{C})$, where $W = (D, P, F)$ is a workflow and $\mathcal{E}$ and $\mathcal{C}$ are mappings of type $\bigcup_{p \in P} \bullet p \times \{p\} \times p \bullet \rightarrow \mathbb{R}_+$, associating a differential privacy and sensitivity value (respectively) to an output produced by a processing node, relative to an input of this processing node.

For example, a privacy-enhanced version of the workflow shown in Fig. 2 is shown in Fig. 3. In the figure, we use $\epsilon_A[x_1, x_3] = 0.2$ to denote the tuple $(x_1, A, x_3, 0.2) \in \mathcal{E}$, meaning that performing A is ϵ-differential private with $\epsilon = 0.2$, when processing x_1 as input and producing x_3. Similarly, $c_A[x_1, x_3] = 0.4$ is used to denoted the tuple $(x_1, A, x_3, 0.4) \subset \mathcal{C}$, which means that A takes as input x_1 and produces x_3 with a sensitivity of 0.4.

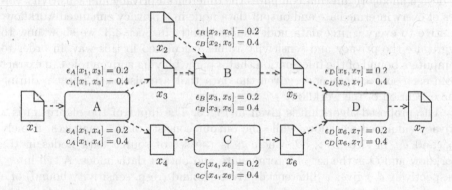

Fig. 3. Example of a privacy-enhanced workflow

A workflow $W = (D, P, F)$ is *interpreted* in the following manner. For each $d \in D$, there is a set X_d and a metric d_d on $\mathcal{D}(X_d)$. For each $p \in P$ and each $d \in p\bullet$, there is a mapping $f_{p \rightarrow d} : \prod_{d' \in \bullet p} X_{d'} \rightarrow \mathcal{D}(X_d)$, which can be lifted to $\overline{f}_{p \rightarrow d} : \prod_{d' \in \bullet p} \mathcal{D}(X_{d'}) \rightarrow \mathcal{D}(X_d)$.

Let $S \subseteq D$ be the set of all source data nodes d, such that $\bullet d = \emptyset$. For each $d \in D$, an interpretation of W defines a mapping $[\![W]\!]_d : \prod_{d' \in S} \mathcal{D}(X_{d'}) \rightarrow \mathcal{D}(X_d)$ as follows. Let $\chi_{d'} \in \mathcal{D}(X_{d'})$ for each $d' \in S$ and let $\mathbf{X}$ be the tuple $(\chi_{d'})_{d' \in S}$. Then

$$[\![W]\!]_d(\mathbf{X}) = \begin{cases} \chi_d, & \text{if } d \in S \\ f_{p \rightarrow d}(([\![W]\!]_{d''}(\mathbf{X}))_{d'' \in \bullet p}), & \text{otherwise, where } \{p\} = \bullet d. \end{cases}$$

The mappings $[\![W]\!]_d$ are well-defined due to the acyclicity of W.

The annotations of a privacy-enhanced workflow $(W, \mathcal{E}, \mathcal{C})$ *match* the interpretation of W if for all $p \in P$, $d' \in \bullet p$ and $d \in p\bullet$,

- the sensitivity of $\overline{f}_{p \to d}$ in its argument "d'" is $c_p[d', d]$ with respect to the distances $\mathrm{d}_{d'}$ on $\mathcal{D}(X_{d'})$ and d_d on $\mathcal{D}(X_d)$;
- the sensitivity of $\overline{f}_{p \to d}$ in its argument "d'" is $\epsilon_p[d', d]$ with respect to the distances $\mathrm{d}_{d'}$ on $\mathcal{D}(X_{d'})$ and d_{dp} on $\mathcal{D}(X_d)$.

These requirements talk about metrics over the sets of probability distributions $\mathcal{D}(X_d)$, and sensitivities of lifted mappings in terms of these metrics. In Appendix A we discuss, how metrics on sets X_d can be lifted to probability distributions and what should the sensitivities of the original mappings $f_{p \to d}$ be.

4.2   Data Node-Based Analysis of Workflows

As stated in Sect. 1, we are interested in computing upper bounds of the information disclosed when data nodes are accessed by a user playing a given role in the process. In order to do so, we leverage the concepts and definitions of Sect. 3 to design an algorithm that computes the differential privacy and sensitivity values of every intermediate and output data node in a privacy enhanced workflow, relative to every source data node. Subsequently in Sect. 4.3, we show how to aggregate the privacy and sensitivity values calculated in this way, in order to compute a bound of the information that a party playing a given role can extract from each source data node, given the data that are disclosed to them during one execution of the workflow.

The proposed algorithm is given in Fig. 3. The input of the algorithm is a privacy-enhanced workflow, while the output consists of two matrices, namely d_{dp} and d_c, of size $|S| \times |O|$ where S is the set of source data nodes in the workflow and O is the set of intermediate and output data nodes. A cell in d_{dp} (respectively d_c) gives a differential privacy bound (resp. sensitivity bound) of a given intermediate or output data node o relative to a source data node s. The main idea of the algorithm is to iterate over the processing nodes in the workflow in topological order (which requires that the workflow is well-formed and thus acyclic). At each step, we compute the value of $d_{dp}[s, o]$ and $d_c[s, o]$ for each output o of the current processing node p, using the previously computed values for the input data nodes of p, as well as the formulas for composing sensitivity values given in Propositions 1, 2 and 3 of Deliverable D1.1 and existing formulas for composition of ϵ−differentially private information release mechanisms.

Example 1. We use the example in Fig. 3 to illustrate Algorithm 1. To this end, we consider the topological order $[A, B, C, D]$ of processing nodes.[3]

During the first iteration, in line 1 the algorithm sets p to the processing node A. In line 2, the algorithm iteratively selects a source data node (i.e. $s \in D : | \bullet s| = 0$) and one successor of p such that the latter is reachable from the selected source node. The first iteration of the inner loop then processes the pair $s = x_1$ and $o = x_3$. Since x_1 is a direct predecessor of A the algorithm will perform lines 4–5. As a result, we have that $d_{dp}[x_1, x_3] = \epsilon_A[x_1, x_3] = 0.2$ and

---

[3] Note that there exists another topological order of the processing nodes of the example, namely $[A, C, B, D]$. Either one would produce the same output matrices.

Algorithm 1. Differential privacy of a workflow

Data: A well-formed *workflow* (W, S), with $W = (D, P, F)$

Result: The matrices d_{dp} and d_c

1 **foreach** processing node $p \in P$ in topological order **do**

2   **foreach** $s \in D, o \in p\bullet : |\bullet s| = 0 \wedge (s, o) \in F^+$ **do**

3   **if** $s \in \bullet p$ **then**

4              $d_{dp}[s, o] = \epsilon_p[s, o]$

5              $d_c[s, o] = c_p[s, o]$

6   **else**

7              $d_{dp}[s, o] = \sum_{i \in \bullet p:(s,i) \in F^+} \min\left(d_{dp}[s, i], \; d_c[s, i] \cdot \epsilon_p[i, o]\right)$

8              $d_c[s, o] = \sum_{i \in \bullet p:(s,i) \in F^+} \left(d_c[s, i] \cdot c_p[i, o]\right)$

9   **end**

10   **end**

11 **end**

12 **return** d_{dp}, d_c

$d_c[x_1, x_3] = c_A[x_1, x_3] = 0.4$. The second iteration of the inner loop, in turn, will process the pair $s = x_1$ and $o = x_4$. The latter will result in $d_{dp}[x_1, x_4] = \epsilon_A[x_1, x_4] = 0.2$ and $d_c[x_1, x_4] = c_A[x_1, x_4] = 0.4$. This will complete the first iteration of the outer loop because none of the successors of A is reachable from x_2. The following matrices summarize the outcome of the first iteration:

	x_3	x_4	x_5	x_6	x_7
x_1	$\epsilon_A[x_1, x_3] = 0.2$	$\epsilon_A[x_1, x_4] = 0.2$			
x_2					

d_{dp}

	x_3	x_4	x_5	x_6	x_7
x_1	$c_A[x_1, x_3] = 0.4$	$c_A[x_1, x_4] = 0.4$			
x_2					

d_c

In the second iteration, the algorithm sets p to the processing node B (line 1). The inner loop first computes the values for source node x_1 and the only successor of B, that is x_5. This time, the algorithm executes lines 7–8, because x_1 is not a direct predecessor of B. Note that x_3 is the only direct predecessor of B which is reachable from x_1 and, as a result, there is only one term in the summation of line 7. Therefore, in line 7 we have that $d_{dp}[x_1, x_5] = \min(d_{dp}[x_1, x_3], d_c[x_1, x_3] \cdot \epsilon_B[x_3, x_5]) = \min(0.2, 0.4 \cdot 0.2) = 0.08$ and in line 8 $d_c[x_1, x_5] = d_c[x_1, x_3] \cdot c_B[x_3, x_5] = 0.4 \cdot 0.4 = 0.16$. In the second iteration of the inner loop, the algorithm computes the values associated to the source node x_2 and the only successor of b, that is x_5. Since x_2 is direct predecessor of B, the algorithm sets $d_{dp}[x_2, x_5] = \epsilon_B[x_2, x_5] = 0.2$ and $d_c[x_2, x_5] = c_B[x_2, x_5] = 0.4$.

The third iteration selects $p = C$ and proceeds in a similar way as for the second iteration. The following matrices summarize the values computed at the end of this iteration.

	x_3	x_4	x_5	x_6	x_7
x_1	0.2	0.2	0.08	0.08	
x_2			0.2		

d_{dp}

	x_3	x_4	x_5	x_6	x_7
x_1	0.4	0.4	0.16	0.16	
x_2			0.4		

d_c

In the final iteration, the algorithm computes the values by selecting p to be the processing node D. In the inner loop, the algorithm will first select the source node x_1. Note that D has x_7 as its only successor. However, there are two direct predecessors of D, namely x_5 and x_6. Therefore the computation of d_{dp} involves the summation of the values that come from x_5 and x_6. Thus, we have that:

$$d_{dp}[x_1, x_7] = \min\left(d_{dp}[x_1, x_5], d_c[x_1, x_5] \cdot \epsilon_D[x_5, x_7]\right) +$$
$$\min\left(d_{dp}[x_1, x_6], d_c[x_1, x_6] \cdot \epsilon_D[x_6, x_7]\right)$$
$$= \min\left(0.08, 0.16 \cdot 0.2\right) + \min\left(0.08, 0.16 \cdot 0.2\right)$$
$$= 0.064$$

and

$$d_c[x_1, x_7] = \left(d_c[x_1, x_5] \cdot c_D[x_5, x_7]\right) + \left(d_c[x_1, x_6] \cdot c_D[x_6, x_7]\right)$$
$$= \left(0.16 \cdot 0.4\right) + \left(0.16 \cdot 0.4\right)$$
$$= 0.128$$

In the final iteration of the inner loop, the algorithm computes the values for $s = x_2$ and $o = x_5$. In this case however, there is only one term in the summation. Therefore, $d_{dp}[x_2, x_7] = \min\left(d_{dp}[x_2, x_5], d_c[x_2, x_5] \cdot \epsilon_D[x_5, x_7]\right) = \min\left(0.2, 0.4 \cdot 0.2\right) = 0.08$. Finally, $d_c[x_2, x_7] = d_c[x_2, x_5] \cdot c_D[x_5, x_7] = 0.4 \cdot 0.4 = 0.16$. The following matrices summarize the outcome of the algorithm.

	x_3	x_4	x_5	x_6	x_7
x_1	0.2	0.2	0.08	0.08	0.064
x_2			0.2		0.16

d_{dp}

	x_3	x_4	x_5	x_6	x_7
x_1	0.4	0.4	0.16	0.16	0.128
x_2			0.4		0.08

d_c

The correctness of Algorithm 1 is established by the following theorem.

Theorem 1. *Let $(W, \mathcal{E}, \mathcal{C})$ be a privacy-enhanced workflow with $W = (D, P, F)$. Let W have an interpretation that matches the annotations $\mathcal{E}$ and $\mathcal{C}$. Let $x \in S$ and $y \in O$. Let the matrices d_{dp} and d_c be computed by Algorithm 1 from $(W, \mathcal{E}, \mathcal{C})$. Then $[\![W]\!]_y$ is $d_{dp}[x, y]$-differentially private and $d_c[x, y]$-sensitive in its argument "x" according to the distances d_x on $\mathcal{D}(X_x)$ and d_y on $\mathcal{D}(X_y)$.*

Proof. The theorem is proved by induction over the data nodes of W, taken in topological order. First we amend d_{dp} and d_c with columns corresponding to data nodes in S, defining $d_c(x, x) = 1$ and $d_{dp}(x, x) = \infty$ for all $x \in S$, as well as $d_c(x, y) = d_{dp}(x, y) = 0$ for all $x, y \in S$ with $x \neq y$. We now proceed with the induction.

Base case: $y \in S$. Then $[\![W]\!]_y$ takes the component "y" from its tuple of arguments. The values $d_{dp}(x,y)$ and $d_c(x,y)$ describe the sensitivity and differential privacy of the protection.

Induction step: let $\{p\} = \bullet y$ and let the differential privacy and sensitivity claims hold for all $[\![W]\!]_{y'}$, where $y' \in \bullet p$. We note that $f_{p \to y}$ is 1-sensitive for all its inputs, if the distance on both the input and the output is d_{dp}. The reason for this is, that no post-processing can degrade the privacy level of a differentially private mapping.

Let $C_{y'}$ be the sensitivity of $[\![W]\!]_{y'}$ in its argument "x". By induction hypothesis, $C_{y'} \leq d_c[x,y']$. Proposition 2 now gives us that the sensitivity of $[\![W]\!]_y$ in its argument "x" is at most $\sum_{y' \in \bullet p} C_{y'} \cdot c_p[y',y] \leq \sum_{y' \in \bullet p} d_c[x,y'] \cdot c_p[y',y] = d_c[x,y]$ by Algorithm 1.

Similarly, let $E_{y'}$ be differential privacy level of $[\![W]\!]_{y'}$ in its argument "x"; by induction hypothesis, $E_{y'} \leq d_{dp}[x,y']$. On each $\mathcal{D}(X_{y'})$, we may consider either the distance $\mathsf{d}_{y'}$ or d_{dp}. The sensitivity of $[\![W]\!]_{y'}$ (in argument "x") is $\mathbf{c}^1_{y'} = C_{y'}$ according to $\mathsf{d}_{y'}$ or $\mathbf{c}^0_{y'} = E_{y'}$ according to d_{dp}. The sensitivity of $f_{p \to y}$ in argument "y'" according to the distance $\mathsf{d}_{y'}$ [resp. d_{dp}] on $\mathcal{D}(X_{y'})$ and the distance d_{dp} on $\mathcal{D}(X_y)$ is $\epsilon^1_{y'} = \epsilon_p[y',y]$ [resp $\epsilon^0_{y'} = 1$]. Let $b[y'] \in \{0,1\}$ for each $y' \subset \bullet p$. According to Proposition 2, the differential privacy of $[\![W]\!]_y$ is $\sum_{y' \subset \bullet p} \mathbf{c}^{b[y']}_{y'} \cdot \epsilon^{b[y']}_{y'}$, obtained by considering the distance $\mathsf{d}_{y'}$ (if $b[y'] = 1$) or d_{dp} (if $b[y'] = 0$) on $\mathcal{D}(X_{y'})$. This bound for differential privacy holds for any choice of bits $b[y']$. Hence the differential privacy of $[\![W]\!]_y$ is

$$\min_{\forall y' \in \bullet p : b[y'] \in \{0,1\}} \sum_{y' \in \bullet p} \mathbf{c}^{b[y']}_{y'} \cdot \epsilon^{b[y']}_{y'} = \sum_{y' \in \bullet p} \min(\mathbf{c}^0_{y'} \cdot \epsilon^0_{y'}, \mathbf{c}^1_{y'} \cdot \epsilon^1_{y'}) =$$

$$\sum_{y' \in \bullet p} \min(E_{y'} \cdot 1, C_{y'} \cdot c_p[y',y]) \leq \sum_{y' \in \bullet p} \min(d_{dp}[x,y'], d_c[x,y'] \cdot c_p[y',y]) = d_{dp}[x,y]$$

by Algorithm 1.

4.3   Role-Based Privacy Analysis of Workflows

So far, we have considered workflows without a notion of *parties* to whom data is disclosed. To capture this latter aspect, we extend the notion of privacy-enhanced workflow with a disclosure relation $Disc \subseteq D \times R$, such that $Disc(n,r)$ denotes the fact that data node n is disclosed to role r. We assume here a classical role-based access control model, entailing that all users who play role r are able to access all data nodes disclosed to role r.

Given the matrices d_{dp} and d_c computed from a privacy-enhanced workflow W and given the relation $Disc$ capturing the disclosure of data nodes to roles, we can now compute a differential privacy bound $\epsilon_r(s)$ of the information that a given role r can extract from a given source data node s – i.e. how much a party playing a given role can learn about individual records of a given input s of W:

$$\epsilon_r(s) = \sum_{(n,r)\, \in\, Disc\, :\, (s,n)\, \in\, F^+} d_{dp}[s,n] \tag{2}$$

Example 2. Given the matrices computed in the previous example:

	x_3	x_4	x_5	x_6	x_7
x_1	0.2	0.2	0.08	0.08	0.064
x_2			0.2		0.16

d_{dp}

	x_3	x_4	x_5	x_6	x_7
x_1	0.4	0.4	0.16	0.16	0.128
x_2			0.4		0.08

d_c

we can compute the differential privacy guarantee with respect to data node x_1 that can be made for a party playing a role r that has access to both data nodes x_5 and x_6 in the workflow shown in Fig. 3:

$$\epsilon_r(x_1) = d_{dp}[x_1, x_5] + d_{dp}[x_1, x_6]$$
$$= 0.08 + 0.08 = 0.16$$

In Eq. 2, we sum up the ϵ values calculated for each intermediate/output data node that is disclosed to role r. This is a worst-case bound, applicable in the case of "sequential composition" of differentially private release mechanisms. The underpinning assumption is that if a role has access to two data nodes n_1 and n_2 produced from a given source data node s via two different paths in the workflow, these two paths use the same or overlapping parts of data source s. If this is not the case, meaning that n_1 and n_2 derive from independent parts of s, a tighter bound may be applied – in the best case $\max(d_{dp}[s, n_1], d_{dp}[s, n_2])$ instead of $d_{dp}[s, n_1] + d_{dp}[s, n_2]$, based on existing results for so-called "parallel composition" of differentially private release mechanisms. Hence, if we annotated a privacy-enhanced workflow with additional metadata capturing independence relations between multiple accesses to the same data source, we could refine the calculation of the differential privacy budgets. Investigating this refinement is a direction for future work.

5   Related Work

Differential privacy has been widely studied in the context of program analysis, using e.g. types [10] or theorem proving [2]. These techniques allow one to reason about the differential privacy of the output of a program relative to its input. In a similar vein, techniques have been proposed to analyze sensitivity and differential privacy for database queries expressed in SQL-like languages (e.g. PINQ) [12]. In this latter work, the aim is to ensure that the output of a given query is differentially private with respect to the input tables, for a given privacy budget. Again, these techniques are geared at analyzing that the output of a processing node is differentially private with respect to its input. In this respect, these proposals are complementary to ours: They can be used to analyze the sensitivity and differential privacy of each data processing node in a workflow.

Perhaps the closest work to ours is Featherweight PINQ [7]. This latter work defines a calculus that can be used to determine the sensitivity of parallel and sequential compositions of queries defined in a workflow-like notation. However, it does not provide a framework that combines sensitivity and differential privacy under the same roof as we do in our proposal.

Our work is also related to *d-privacy* [14]: a generalization of differential privacy that turns any metric on the set of possible datasets into a composable privacy metric, of which differentially privacy is a special case. In this paper, we build upon these and related ideas in [4,8] in order to define new composition rules for processing nodes with differentially-private release mechanism, specifically rules that combine sensitivity and differential privacy in a way that allows us to calculate differentially private bounds when a differentially-private output of a node is fed as input to another differentially-private node.

Previous work on privacy analysis of business processes [1,9] relies on Petri net reachability analysis and model checking to detect data objects that are declassified to unauthorized parties either in full or in part. These approaches adopt a multi-level security model, wherein the objects and subjects of the system are divided in security levels. The goal of these techniques is to identify cases where information from an object of a higher security level is copied to an object with lower security level. However such techniques are Boolean: they detect potential leakages but fail to quantify them, which is the goal of the present paper.

The workflow notation employed in this paper is conceptually similar to graphical workflow notations used in data warehousing [11] – where they are referred to as Extract Transform-Load (ETL) workflows – and also bears resemblances with data analytics workflow notations such as the one embodied in the popular KNIME toolset [3]. The results presented in this paper can potentially be applied to analyze workflows in these related notations.

6   Conclusion

To summarize, the main contributions of this paper are:

1. Theoretical results on sensitivity of composed mappings that generalize well-known composition theorems of differential privacy and allow us to calculate differential privacy bounds in the case where the differentially-private output of a function is used as input to another differentially-private function.
2. A notion of privacy-enhanced workflow where tasks (processing nodes) transform input objects into output objects using differentially private mechanisms and each intermediate or output object is disclosed to one or more roles.
3. An algorithm that given a data processing workflow and given the sensitivity and differential privacy leakage of each processing node in the workflow, estimates the differential privacy leakage generated by the disclosure of data to each role involved in the workflow.

The proposed analysis technique has been implemented in a tool called Pleak.[4] Pleak allows one to: (i) model a process using the standard Business

---

[4] The tool is available at http://pleak.io/ for research purposes.

Process Model and Notation (BPMN); (ii) annotate the elements of the process model with sensitivity and differential privacy metadata; and (iii) obtain a table stating the differential privacy budget consumed by each role (lane or pool) in the process relative to each input data collection. Internally, the tool extracts a data processing workflow from the BPMN model, and applies the technique presented above. At present, only a subset of BPMN is supported, comprising tasks, sequence flows, parallel gateways, data objects, lanes and pools. Also, the input process models are assumed to be acyclic. These restrictions are meant to ensure the BPMN model can be transformed into a data processing workflow.

In future, we will extend the notion of data processing workflow in order to lift some of the restrictions imposed so far, particularly the restriction that data processing workflows do not contain conditional branching nor cycles. To this end, we need to extend the theoretical foundation to reason about conditional branches in a differentially private computation. Also, as stated in Sect. 4.3, we plan to enhance the workflow notation to capture independence relations between multiple accesses to the same data source, so as to calculate tighter bounds for differential privacy budgets when multiple computations access independent parts of the same data collection (e.g. distinct sets of attributes).

Acknowledgments. This work is funded by DARPA's "Brandeis" programme.

A   Lifting Distances to Probability Distributions

To interpret a privacy-enhanced DP-workflow $(W, \mathcal{E}, \mathcal{C})$ (where $W = (D, P, F)$), we have to give metrics d_d on $\mathcal{D}(X_d)$ for each $d \in D$. Moreover, for the interpretation to be matched by the annotations, the mappings $\overline{f}_{p \to d}$ between these probability distributions must have the sensitivities given by $\mathcal{E}$ and $\mathcal{C}$. It may be more natural to assume that the interpretation gives us metrics on X_d, not on $\mathcal{D}(X_d)$. It is also more natural to require the mappings $f_{p \to d}$ to have a certain sensitivity.

We thus define that a *pre-interpretation* consists of sets X_d for each $d \in D$ together with a metric $\mathsf{d}_d^{\flat}$ on it, as well as the mappings $f_{p \to d}$ for each $p \in P$ and $d \in p\bullet$. We have to specify what kind of interpretation it generates, and when to the annotations $\mathcal{E}, \mathcal{C}$ match the pre-interpretation. The key for this is to specify the metric d_d on $\mathcal{D}(X_d)$.

Let X be a set and d_X a metric on it. It turns out that the following definition of a metric $d_X^{\#}$ on $\mathcal{D}(X)$ is a suitable one. Let $\chi, \chi' \in \mathcal{D}(X)$. Then

$$d_X^{\#}(\chi, \chi') = \inf_{\psi \in \chi \otimes \chi'} \sup_{(x,x') \in \mathrm{supp}(\psi)} d_X(x, x'). \tag{A.1}$$

The proposed metric $d_X^{\#}$ can be seen as a kind of "worst-case" earth mover's distance (or Wasserstein metric). In the "usual" earth mover's distance, one would take the average over ψ, not the supremum over $\mathrm{supp}(\psi)$.

The suitability of the construction (A.1) is given by the following two propositions. Note that the first of them would not hold for the "usual" earth mover's distance.

Proposition 3. *Let $f : X \to \mathcal{D}(Y)$ be ε-sensitive according to the distance d_X on X and distance d_{dp} on $\mathcal{D}(Y)$. Then the lifting $\overline{f} : \mathcal{D}(X) \to \mathcal{D}(Y)$ is ε-sensitive according to the distance $d_X^\#$ on $\mathcal{D}(X)$ and d_{dp} on $\mathcal{D}(Y)$.*

Proof. Let $\chi, \chi' \in \mathcal{D}(X)$, $\psi \in \chi \otimes \chi'$ and $y \in Y$. Then

$$
\Pr[\overline{f}(\chi) = y] = \sum_{x \in X} \chi(x) \cdot \Pr[f(x) = y] = \sum_{x,x' \in X} \psi(x, x') \cdot \Pr[f(x) = y] \leq
$$

$$
\sum_{x,x' \in X} \psi(x, x') \cdot e^{\varepsilon \cdot d_X(x,x')} \Pr[f(x') = y] \leq
$$

$$
\sum_{x,x' \in X} \psi(x, x') \cdot e^{\sup_{x \in \mathrm{supp}(\psi(\cdot,x'))} \varepsilon \cdot d_X(x,x')} \Pr[f(x') = y] =
$$

$$
\sum_{x' \in X} \chi'(x') \cdot e^{\sup_{x \in \mathrm{supp}(\psi(\cdot,x'))} \varepsilon \cdot d_X(x,x')} \Pr[f(x') = y] \leq
$$

$$
e^{\sup_{x,x' \subseteq \mathrm{supp}(\psi)} \varepsilon \cdot d_X(x,x')} \cdot \sum_{x' \in X} \chi'(x') \cdot \Pr[f(x') = y] =
$$

$$
e^{\sup_{x,x' \in \mathrm{supp}(\psi)} \varepsilon \cdot d_X(x,x')} \cdot \Pr[\overline{f}(\chi') = y],
$$

where $\mathrm{supp}(\psi(\ ,x'))$ denotes the set of all $x \in X$, such that $\psi(x, x') > 0$. We obtain

$$
d_{\mathrm{dp}}(\overline{f}(\chi), \overline{f}(\chi')) = \sup_{y \in Y} \left| \ln \frac{\Pr[\overline{f}(\chi') = y]}{\Pr[\overline{f}(\chi) = y]} \right| \leq
$$

$$
\inf_{\psi \in \chi \otimes \chi'} \sup_{x,x' \in \mathrm{supp}(\psi)} \varepsilon \cdot d_X(x, x') = \varepsilon \cdot d_X^\#(\chi, \chi').
$$

Proposition 4. *Let $f : X \to \mathcal{D}(Y)$ be c-sensitive according to the distance d_X on X and distance $d_Y^\#$ on $\mathcal{D}(Y)$, where $d_Y^\#$ is constructed from some distance d_Y on Y according to (A.1). Then $\overline{f} : \mathcal{D}(X) \to \mathcal{D}(Y)$ is c-sensitive according to the distance $d_X^\#$ on $\mathcal{D}(X)$ and $d_Y^\#$ on $\mathcal{D}(Y)$.*

Proof. Let $\chi, \chi' \in \mathcal{D}(X)$. Define $\mathbf{F}$ as the following set of mappings of type $X \times X \to \mathcal{D}(Y \times Y)$:

$$
\mathbf{F} = \{ \xi \mid \forall x, x' \in X : \xi(x, x') \in f(x) \otimes f(x') \}.
$$

Also consider the set $\Phi \subseteq \mathcal{D}(Y \times Y)$, defined as follows:

$$
\Phi = \{ \sum_{x,x' \in X} \psi(x, x') \cdot \xi(x, x') \mid \psi \in \chi \otimes \chi', \xi \in \mathbf{F} \}.
$$

In the definition of Φ, we take the averages over $\xi(x, x')$ with the weights given by $\psi(x, x')$. We have $\Phi \subseteq \overline{f}(\chi) \otimes \overline{f}(\chi')$ because the first [resp. second] projection of any element of Φ is $\overline{f}(\chi)$ [resp. $\overline{f}(\chi')$]. We now have

$$d_Y^\#(\overline{f}(\chi), \overline{f}(\chi')) = \inf_{\phi \in \overline{f}(\chi) \otimes \overline{f}(\chi')} \sup_{(y,y') \in \mathrm{supp}(\phi)} d_Y(y,y') \leq$$

$$\inf_{\phi \in \Phi} \sup_{(y,y') \in \mathrm{supp}(\phi)} d_Y(y,y') = \inf_{\psi \in \chi \otimes \chi'} \inf_{\xi \in \mathbf{F}} \sup_{(x,x') \in \mathrm{supp}(\psi)} \sup_{(y,y') \in \mathrm{supp}(\xi(x,x'))} d_Y(y,y') =$$

$$\inf_{\psi \in \chi \otimes \chi'} \sup_{(x,x') \in \mathrm{supp}(\psi)} \inf_{\phi \in f(x) \otimes f(x')} \sup_{(y,y') \in \mathrm{supp}(\phi)} d_Y(y,y') =$$

$$\inf_{\psi \in \chi \otimes \chi'} \sup_{(x,x') \in \mathrm{supp}(\psi)} d_Y^\#(f(x), f(x')) \leq \inf_{\psi \in \chi \otimes \chi'} \sup_{(x,x') \in \mathrm{supp}(\psi)} c \cdot d_X(x,x') = c \cdot d_X^\#(\chi, \chi')$$

These two propositions tells us how to turn a pre-interpretation of a privacy-enhanced DP-workflow into an interpretation. We define $\mathsf{d}_d = (\mathsf{d}_d^\flat)^\#$ for each $d \in D$. The annotations $\mathcal{E}, \mathcal{C}$ match the pre-interpretation if for all $p \in P$, $d' \in \bullet p$ and $d \in p\bullet$:

- the sensitivity of $f_{p \to d}$ in its argument "d'" is $c_p[d', d]$ with respect to the distances $\mathsf{d}_{d'}^\flat$ on $X_{d'}$ and d_d on $\mathcal{D}(X_d)$;
- the sensitivity of $f_{p \to d}$ in its argument "d'" is $\epsilon_p[d', d]$ with respect to the distances $\mathsf{d}_{d'}^\flat$ on $X_{d'}$ and d_{dp} on $\mathcal{D}(X_d)$.

In this way, the corresponding interpretation is also matched by the annotations.

References

1. Accorsi, R., Lehmann, A., Lohmann, N.: Information leak detection in business process models: theory, application, and tool support. Inf. Syst. **47**, 244–257 (2015)
2. Barthe, G., Köpf, B., Olmedo, F., Béguelin, S.Z.: Probabilistic relational reasoning for differential privacy. ACM Trans. Program. Lang. Syst. **35**(3), 9 (2013)
3. Berthold, M.R., Nicolas, C., Dill, F., Gabriel, T.R., Kötter, T., Meinl, T., Ohl, P., Thiel, K., Wiswedel, B.: KNIME - the Konstanz Information Miner: version 2.0 and beyond. SIGKDD Explor. **11**(1), 26–31 (2009)
4. Chatzikokolakis, K., Andrés, M.E., Bordenabe, N.E., Palamidessi, C.: Broadening the scope of differential privacy using metrics. In: De Cristofaro, E., Wright, M. (eds.) PETS 2013. LNCS, vol. 7981, pp. 82–102. Springer, Heidelberg (2013)
5. Dwork, C.: Differential privacy. In: Bugliesi, M., Preneel, B., Sassone, V., Wegener, I. (eds.) ICALP 2006. LNCS, vol. 4052, pp. 1–12. Springer, Heidelberg (2006)
6. Dwork, C., McSherry, F., Nissim, K., Smith, A.: Calibrating noise to sensitivity in private data analysis. In: Halevi, S., Rabin, T. (eds.) TCC 2006. LNCS, vol. 3876, pp. 265–284. Springer, Heidelberg (2006)
7. Ebadi, H., Sands, D.: Featherweight PINQ (2015). CoRR arXiv:1505.02642
8. ElSalamouny, E., Chatzikokolakis, K., Palamidessi, C.: Generalized differential privacy: regions of priors that admit robust optimal mechanisms. In: van Breugel, F., Kashefi, E., Palamidessi, C., Rutten, J. (eds.) Horizons of the Mind. LNCS, vol. 8464, pp. 292–318. Springer, Heidelberg (2014)
9. Frau, S., Gorrieri, R., Ferigato, C.: Petri net security checker: structural non-interference at work. In: Degano, P., Guttman, J., Martinelli, F. (eds.) FAST 2008. LNCS, vol. 5491, pp. 210–225. Springer, Heidelberg (2009)

10. Gaboardi, M., Haeberlen, A., Hsu, J., Narayan, A., Pierce, B.C.: Linear dependent types for differential privacy. In: Giacobazzi, R., Cousot, R. (eds.) The 40th Annual ACM SIGPLAN-SIGACT Symposium on Principles of Programming Languages, POPL 2013, Rome, Italy, 23–25 January 2013, pp. 357–370. ACM (2013)
11. Kimball, R., Reeves, L., Thornthwaite, W., Ross, M., Thornwaite, W.: The Data Warehouse Lifecycle Toolkit: Expert Methods for Designing, Developing and Deploying Data Warehouses, 1st edn. Wiley, New York (1998)
12. McSherry, F.: Privacy integrated queries: an extensible platform for privacy-preserving data analysis. In: Çetintemel, U., Zdonik, S.B., Kossmann, D., Tatbul, N. (eds.) Proceedings of ACM SIGMOD International Conference on Management of Data, SIGMOD, Providence, Rhode Island, USA, 29th June–2nd July 2009, pp. 19–30. ACM (2009)
13. Object Management Group: Business Process Model and Notation (BPMN) Version 2.0 (2011)
14. Reed, J., Pierce, B.C.: Distance makes the types grow stronger: a calculus for differential privacy. In: Hudak, P., Weirich, S. (eds.) Proceeding of 15th ACM SIGPLAN International Conference on Functional Programming, ICFP 2010, Baltimore, Maryland, USA, 27–29 September 2010, pp. 157–168. ACM (2010)

Bridging Two Worlds: Reconciling Practical Risk Assessment Methodologies with Theory of Attack Trees

Olga Gadyatskaya[1]([✉]), Carlo Harpes[2], Sjouke Mauw[1], Cédric Muller[1,2], and Steve Muller[2]

[1] SnT, University of Luxembourg, Luxembourg, Luxembourg
{olga.gadyatskaya,sjouke.mauw}@uni.lu, cedric.muller.001@student.uni.lu
[2] itrust Consulting, Niederanven, Luxembourg
{harpes,cedric.muller,steve.muller}@itrust.lu

Abstract. Security risk treatment often requires a complex cost-benefit analysis to be carried out in order to select countermeasures that optimally reduce risks while having minimal costs. According to ISO/IEC 27001, risk treatment relies on catalogues of countermeasures, and the analysts are expected to estimate the residual risks. At the same time, recent advancements in attack tree theory provide elegant solutions to this optimization problem. In this paper we propose to bridge the gap between these two worlds by introducing optimal countermeasure selection problem on attack-defense trees into the TRICK security risk assessment methodology.

1 Introduction

Recent attacks, such as Ashley Madison, Sony and Target, are well-known to many of us. However, it is not only large or famous organizations that are targeted by cyber criminals. Any company can be attacked, and companies have to respond to this huge threat landscape by improving their security protection. Nowadays the ability to better identify and prioritize security risks, and to detect and mitigate incidents becomes *critical*. Companies need to look for the means to pinpoint and quantify security gaps and to eliminate them by introducing new security controls. Usually controls are selected following some established guidelines. There exist *generic* security guidelines, e.g., IT-Grundschutz Catalogues [4], ISO/IEC 27002 [13], NIST 800-53 [19], and *domain-specific* ones. Examples of the latter are PCI DSS [22] in the banking domain, the controls catalogue [6] in the air traffic management domain, ISO 27799 [10] for health informatics, ISO 27019 [14] for the energy utility industry. Furthermore, controls can be also identified by the interested parties and analysts in brainstorming [24].

The research leading to the results presented in this work received funding from the European Commission's Seventh Framework Programme (FP7/2007–2013) under grant agreement number 318003 (TREsPASS) and Fonds National de la Recherche Luxembourg under the grant C13/IS/5809105 (ADT2P).

B. Kordy et al. (Eds.): GraMSec 2016, LNCS 9987, pp. 80–93, 2016.
DOI: 10.1007/978-3-319-46263-9_5

On the other hand, in the academic world there exist many techniques and tools to select countermeasures in an optimal way. These techniques can be roughly classified as more generic (e.g., optimal countermeasure selection on attack trees [2,25]), or more domain-specific (for example, network hardening techniques on attack graphs [1]).

These two worlds focused on the same problem of *countermeasure selection* rarely engage with each other, one of the reasons being that industrial risk treatment practices are entangled with many other practices and processes in the company (governance and compliance, but also business operations), while academic solutions tend to be more isolated and focused on particular aspects. Furthermore, design of new risk assessment methods generally follows the requirements and guidelines imposed by relevant standardization and regulation bodies [23], i.e., ISO 27001 [12] and NIST Cybersecurity Framework [20]. Academic solutions need to be introduced into risk management methodologies *on top* of these guidelines. In this position paper we propose to bridge the two worlds of practical risk management and theoretical results on optimal security control selection in attack trees. As the security risk assessment method we apply the TRICK Service framework developed and used in Luxembourg. We consider to bridge this practical assessment process with an academic result concerning the optimal countermeasure selection problem on attack trees, which is an instance of approaches proposed by Roy et al. [25] and Aslanyan and Nielson [2].

The paper is structured as follows. We give an outline of the TRICK Service in Sect. 2, and present a background on attack tree theory in Sect. 3. Our proposal for bridging these two domains in the context of optimal selection of countermeasures in risk treatment is presented in Sect. 4. We discuss possible choices for selecting countermeasures in Sect. 5, and we present the optimization problem that we solve for allocation of defensive nodes in attack trees in Sect. 6. We illustrate our current approach on a private cloud use case in Sect. 7. We then overview our next steps and conclude in Sect. 8.

2   The TRICK Service

TRICK Service (**T**ool for **R**isk management of an **I**nformation Security Management System based on a **C**entral **K**nowledge base), developed by itrust consulting in Luxembourg, is a web-based risk assessment and management tool for identification, analysis and estimation of assets, threats, vulnerabilities, risk scenarios and security measures. It helps the analyst to determine a list of security measures to be implemented in order to reduce the impact or the likelihood of possible risk scenarios.

Risk analysis in TRICK starts with establishing the context by collecting information about the type and business processes of the organisation and filling in a table, according to ISO 27005:2011 [11]. This information is used by the analyst to establish the most important assets considering the sector of the organisation.

After the context definition, a brainstorming session identifies assets and risk scenarios in the organisation. Qualitative risk assessment is performed at this stage to allow the analyst to estimate the exposure to identified threats, vulner-abilities and risks. The next step consists in identifying the security measures that are already implemented in the organization, and assessing their current implementation rate and cost, referring to norms, such as ISO/IEC 27002 [13].

The analyst then estimates the *annual loss expectancy* (ALE) of each asset-scenario pair, by multiplying the impact (in euros) that a scenario could have, with the annual expected probability that a scenario could occur on the asset.

A *risk reduction factor* (RRF) parameter is associated to each asset-scenario-countermeasure triple. The RRF is a coefficient that expresses the negative influ-ence of a security control on the ALE generated by the occurrence of a scenario on an asset. For a given security control in relation to a given scenario acting on an asset, its RRF is a value between 0 and 1, where RRF=0 means that the countermeasure is useless, and RRF=1 signifies perfect protection.

Implementation (or partial implementation) of a security control results in an ALE reduction, based on the RRF and the implementation rate. For the sake of simplicity we will not take the implementation rate of a security measure into account, assuming that any countermeasure is fully implemented.

As we have seen from the description, in order to ensure that the overall risk assessment, analysis and treatment process is correct, the analyst needs to come up with a (sufficiently) complete list of scenarios and evaluate their respective probabilities. If scenarios are too generic, it is very challenging to evaluate their probabilities (or occurrence rates). At the same time, for simpler attack steps, e.g., vulnerability exploitation, it might be more easy to evaluate their chances to occur by relying, e.g., on the available statistics in the sector. To better estimate the residual ALE, we proposed to apply the attack tree formalism summarized in the next section.

3   Attack Tree Theory Background

Attack trees [26] are a graphical model useful for threat modelling and risk assessment [18,21]. They are comprehensible to stakeholders with different back-grounds and expertise, and they also enjoy various formal semantics [17] that allow for qualitative and quantitative analysis of attack scenarios. In a typical attack tree, the top node (the root) represents the goal of the attacker. For instance, a possible goal is *entering the system to manipulate the integrity* (risk scenario) *of financial transactions* (asset) *by arranging a money transfer to the attacker* (impact).

The root is *refined* into a set of child nodes that represent the different ways to achieve the goal. An **or**-refinement means that any child is sufficient to achieve the parent goal, and an **and**-refinement states that all children need to be achieved before the parent is achieved. Consequently, each child node can be further refined, until the remaining nodes are *simple* enough and do not require further refinement. These simple attack nodes are also called *atomic* attacks, and they are leaf nodes of the attack tree.

Probability computations on attack trees. For the scope of this paper we assume that all atomic attacks in the tree are *independent*, and that all attack nodes are unique in the tree. Then for two attack leaf nodes x and y that represent independent events, with respective probabilities $\mathbf{Pr}(x)$ and $\mathbf{Pr}(y)$, we can calculate their composed probability by $\mathbf{Pr}(x \wedge y) = \mathbf{Pr}(x)\mathbf{Pr}(y)$; and $\mathbf{Pr}(x \vee y) = \mathbf{Pr}(x) + \mathbf{Pr}(y) - \mathbf{Pr}(x)\mathbf{Pr}(y)$. A bottom-up evaluation can be further continued on intermediate nodes until the probability of the root node of the attack tree at, denoted as $\mathbf{Pr}(at)$, is computed. This evaluation can be done in, e.g., the ADTool [8,15].

Attack-defense trees. Attack trees consider the situation only from the perspective of the attacker. However, the main goal of using attack trees in practice is to systematize threat identification in order to improve risk treatment, i.e. identification of relevant countermeasures. Therefore, extensions of attack trees with defensive nodes emerged as a way to explicitly tackle the security control problem. Notable extensions include defense trees [3], protection trees [5], attack-countermeasure trees [25], and *attack-defense* trees [16]. In this work we focus on attack-defense trees as this formalism integrates attacks and countermeasures in the least restrictive way (i.e., defense nodes can be interleaved with attack nodes, while in other formalisms they are typically only leaf nodes).

The problem of countermeasure selection is not novel in the context of attack trees. Roy et al. considered the problem of optimal countermeasure selection for attack-countermeasure trees in [25], and Aslanyan and Nielson investigated optimal probability-cost balances on attack defense trees in [2]. Both of these works consider a tree with already pre-selected countermeasures, and the solution of the optimization problem is to find the subset of *already placed* countermeasures, such that the probability of attacker's success and the cost of selected controls are minimal (a set of Pareto-efficient solutions is offered in [2]). Our goal is to introduce optimal countermeasure selection akin to [2,25] into the TRICK risk assessment methodology.

4   Proposal for Bridging the Gap

We consider that the analyst who is using TRICK will now express threat scenarios as attack trees, and will perform the subsequent risk treatment steps using these trees.

The ROSI Function. The *return on security investment* (ROSI) function evaluates the investment made into security controls versus the obtained security improvement [9]. The average yearly cost of implementing a set of new countermeasures M (denoted as $\mathbf{cost}(M)$) corresponds to the investment, and the total ALE reduction obtained as a result of implementing these new countermeasures (denoted $\Delta ALE(M)$) corresponds to the yearly gains. Thus, for a set of controls M, $ROSI(M) = \Delta ALE_M - \mathbf{cost}(M)$.

Considering that the **Risk** equals **Impact** multiplied by **Probability** [3], we set the difference in the annual loss expectancy $\Delta ALE(M)$ as the product of the **Impact** times the difference of yearly probability of occurrence without and with implementation of the set of countermeasures M [25].

The probability for the attacker to reach the goal and to implement the threat scenario can be evaluated through probabilities of atomic attack steps, as discussed in Sect. 3. At the same time, the impact of the attack tree (i.e., the impact in case the attacker reaches his/her goal and the threat scenario expressed in the attack tree has occurred) can be estimated independently from the tree. Thus we focus only on probability values and the selection of countermeasures based on how well they can reduce the attack success probability.

We consider that each countermeasure t has a possible effect on each attack node x. This effect is described by an **effectiveness** parameter, $\mathbf{eff}(t, x)$ $\in [0,]$, with $\mathbf{eff}(t, x) = 0$ corresponding to a useless countermeasure for x, and $\mathbf{eff}(t, x) = 1$ defining perfect protection against x.

The effectiveness is defined so that the overall probability of the attack node x mitigated by t, which we denote as x_t, is defined as $\mathbf{Pr}(x_t) = \mathbf{Pr}(x)(1\text{-}\mathbf{eff}(t, x))$. Thus, the higher the effectiveness parameter of the countermeasure in the given context, the lower the resulting probability of attack.

The step of evaluating the security posture by considering already implemented countermeasures in TRICK, can be directly executed on the attack tree. The analyst will now place the existing countermeasures as defense nodes in the attack tree. Computation of probabilities in presence of countermeasures and their effectiveness can be done via the bottom-up evaluation algorithm; just like for attack trees. As a result of this step of considering already existing countermeasures, the analyst will obtain an attack-defense tree adt and will evaluate the overall probability of the considered attack scenario as $\mathbf{Pr}(adt)$. For simplicity, in this paper we consider that the analyst "starts from scratch", i.e., the infrastructure does not have any security controls implemented yet, and the analyst starts from an attack tree.

An important distinction of effectiveness from RRF in the context of TRICK is that RRF measures global influence of the countermeasure on the particular scenario occurring with the asset (i.e., on the whole attack tree), while effectiveness is more localized as it applies to an attack node (sub-scenario) in the attack tree, and reduces the probability of occurrence of only this node. The RRF in the TRICK context could be further defined for a set of countermeasures as a non-linear combination of their effectiveness parameters in the attack-defence tree. Thus the process of creating an attack tree, estimating the effectiveness parameters of available countermeasures, and selecting the optimal subset of countermeasures can in the future serve as a methodology to better estimate RRFs in the TRICK Service.

New Countermeasures From Catalogues. As we have mentioned, the de-facto standard for risk treatment is to use catalogues of appropriate security mechanisms, such as [4,6,13,19,22]. TRICK also implements the catalogue of standard security controls defined by ISO/IEC 27002 [13], and others. Therefore, a straightforward way to implement optimal countermeasure selection is to consider such a catalogue of countermeasures, and to define an optimization problem on an attack-defense tree that maximizes the ROSI function.

Indeed, in practice an organization cannot implement all potential countermeasures, and often even implementation of the most critical security controls needs to be prioritized due to budget restrictions. Therefore, countermeasure selection needs to be guided by the *cost-benefit analysis*, in which we will consider costs of countermeasures versus their respective benefit (how well they can reduce attack probabilities).

5   Choices for Countermeasure Selection

Several choices are possible for selecting countermeasures. In this section we discuss these options in more detail.

Locality/Universality of Countermeasures. A countermeasure can be *local*, i.e., it has effect only on the attack node it has been applied to in the tree. In this case, if t is selected as a countermeasure for the attack node x, then it reduces the probability of occurrence of the sub-tree x, but does not influence the probability of occurrence of other attack nodes. However, this assumption does not preclude t from being selected as a countermeasure at another applicable attack node y, where it can then reduce the probability of occurrence (while inducing also extra cost of a separate countermeasure). This solution will work well for the cases when indeed separate security controls with the same name need to be introduced in different locations of the infrastructure. For instance, if there are two vulnerable doors that can be used by the attacker to get in, we will be able to propose two door locks as separate protection mechanisms.

Yet, if, for example, an attack tree has the attack nodes "infiltrate the network" and "probe the ports", and the countermeasure "firewall" is applicable to both of them, this countermeasure could be selected as a defense node twice in our solution (so the approach could propose to pay twice for the same firewall). Thus, an alternative is to assume countermeasures to be *universal*, meaning that they are applied once to the entire tree and affect all attack nodes, unless the effectiveness of a countermeasure on a given node has been set to zero (in this case this countermeasure is not shown in the tree). It is also possible to consider the combined approach, when some security controls are local, and some – universal.

Unique/Multiple Countermeasures of the Same Type. One option is to consider that each countermeasure can be applied to an attack node at most once. An alternative solution is to allow multiple identical countermeasures to be applied to the same node. Considering that each countermeasure is unique and can be applied at most once allows to avoid trivial solutions when cheap controls are applied several times. Furthermore, for the countermeasures defined in the ISO/IEC 27002 standard, it makes sense to only apply them once in a given context. Yet, certain defensive mechanisms can in fact improve protection if applied multiple times (e.g., several security guards may be better than one, several locks on a door can be better than a single one).

Combinations of Defense Nodes. In general, catalogues suggest multiple counter-measures against a single attack node. However, the semantics of attack-defense trees only allow one single defense node per attack node [16]. To address this limitation, one can aggregate several applicable countermeasures into a meta-defense node for a given attack node. For example, we consider a combination of defense nodes, expressed as an **and**-refinement, to be added to the tree. Con-sidering t and q to be two countermeasures (extension to the general case of k applicable countermeasures is trivial), we can add to the tree the defense node $t \wedge q$, with $\mathbf{eff}(t \wedge q) = 1 - \mathbf{eff}(t)\mathbf{eff}(q)$. Intuitively it means that both t and q simultaneously provide protection, but their effectiveness may not be fully independent. Furthermore, $\mathbf{cost}(t \wedge q) = \mathbf{cost}(t) + \mathbf{cost}(q)$.

Alternatively, meta-defense nodes can be expressed as an **or**-refinement. In this case, considering two applicable security controls t and q, the aggregated meta-defense node $t \vee q$ can be added to the tree, with $\mathbf{eff}(t \vee q) = 1 - (1 - \mathbf{eff}(t)) \cdot (1 - \mathbf{eff}(q))$. Again, $\mathbf{cost}(t \wedge q) = \mathbf{cost}(t) + \mathbf{cost}(q)$. The choice between these two types of aggregated meta-nodes depends on the interpretation one has for the defense nodes in the attack tree [16].

Defense Location-Sensitivity. If one considers countermeasures to be local, then actual position of the countermeasure in the tree becomes an important fac-tor further contributing to the complexity of the considered problem. We can demonstrate that if a countermeasure t is applicable to both attack nodes x and $x \vee y$ (what is very likely for attack trees expressed in natural language), then assigning t to the parent node provides a better reduction of the risk. Indeed, with the countermeasure assigned to the parent node $x \vee y$, $\mathbf{Pr}(c^p(x \vee y, t)) = \mathbf{Pr}(x \vee y)(1 - \mathbf{eff}(t)) = (\mathbf{Pr}(x) + \mathbf{Pr}(y) - \mathbf{Pr}(x)\mathbf{Pr}(y))(1 - \mathbf{eff}(t))$. In case t is allo-cated with the child node x, we have $\mathbf{Pr}(c^p(x, t) \vee y) = \mathbf{Pr}(x)(1 - \mathbf{eff}(t)) + \mathbf{Pr}(y) - \mathbf{Pr}(x)(1 - \mathbf{eff}(t))\mathbf{Pr}(y)$. It is evident that $\mathbf{Pr}(c^p(x \vee y, t)) - \mathbf{Pr}(c^p(x, t) \vee y) = -\mathbf{Pr}(y)\mathbf{eff}(t) \leq 0$, given that $0 \leq \mathbf{eff}(t), \mathbf{Pr}(y) \leq 1$. Therefore, the closer to the root we place a defense, the better it can reduce the overall probability of the considered attack.

We discuss the choices we have made for our implementation and the opti-mization problem to be solved in the following section.

6   Attack Tree Refinement and Optimization Problem

Assumptions Made on Countermeasure Selection. In our current implementa-tion we assume each security control to be universal. Thus, for each attack node x and each countermeasure t, such that t is applicable to x ($\mathbf{eff}(t, x) > 0$), we consider that t can be applied to x as a defense node everywhere it is applica-ble. Furthermore, we consider that each countermeasure, if selected, is applied exactly once. These considerations imply that the total cost of each countermea-sure is not affected by the number of times this countermeasure appears in the attack-defense tree (it is counted only once). We also consider that aggregated meta-nodes are composed by the ∨-refinement.

The process to refine an estimation of probability for an asset-scenario pair and to find the optimal set of countermeasures is as follows.

A. Assess Input Parameters.

1. *Create an attack tree.* Model the step or variant of the attack and describes them in a pure attack tree at that does not contain any defence notes. Estimate the success probability of each leaf node. Let n be the number of attack nodes in the initial attack tree. Let a_j denote the j-th node in this attack tree. The ADTool [8,15] can be used to compute $\mathbf{Pr}(at)$, which is the success probability of the root note; it depends on the attack tree and the probabilities of the leaf nodes.

2. *Identify countermeasures.* Prepare the list of potentially applicable countermeasures from catalogues. Let m is the number of countermeasures in this list. For each countermeasure, estimate the security implementation costs.

3. *Estimate effectiveness values.* Estimate the value of the $(m \times n)$ *effectiveness matrix* $\mathbf{E}$ indicating the effectiveness of a countermeasure i on an attack node j. We define $\mathbf{E}[i,j] = \mathbf{eff}(i\text{-th countermeasure}, j\text{-th attack})$.

B. Solve the Optimization Problem.
A possible solution of the problem is described by $d = (d_1, d_2, ..., d_m)$, an m-tuple indicating for each countermeasure whether each corresponding countermeasure i will be implemented (if $d_i = 1$) or not ($d_i = 0$). The cost of such a solution is given by $\mathbf{cost}(d) = \sum_{i=1}^{m} (d_i \times \mathbf{cost}(\text{countermeasure } i))$.

Remark that we can have meta-defense nodes. Let the meta-defense node t expresses the combined defenses applicable to the node a_k. Then $\mathbf{eff}(t, a_k) = 1 - \prod_{j=1}^{m}(1 - d_j \times \mathbf{E}[j,k])$. In the attack tree language, this defense node is a node consisting of an ∨-refinement of the selected countermeasures ($d_j = 1$ and $\mathbf{E}[j,k] > 0$).

The Return On Security Investment (of the list of selected countermeasures d) is defined as follows.

$$ROSI(d) = \mathbf{Impact} \cdot (\mathbf{Pr}(at) - \mathbf{Pr}(adt_d)) - \mathbf{cost}(d),$$

where the **Impact** is the loss achieved if the attack succeed (i.e., if the root node of the attack tree occurs), adt_d is the new attack-defense tree in which the countermeasures selected by d have been added to the nodes according to the effectiveness matrix $\mathbf{E}$. Notice that adt can be constructed from at, d, and $\mathbf{E}$. The ADTool can be now used to compute $\mathbf{Pr}(adt_d)$.

Our optimisation problem consists in finding the list of the selected countermeasures d that maximizes $ROSI(d)$.

Note that instead of maximizing $ROSI(d)$, we can as well minimize $\mathbf{Impact} \cdot \mathbf{Pr}(adt_d) + \mathbf{cost}(d)$.

Current Implementation. Our current implementation uses a branching algorithm based on multiple parameters. We use a brute-force algorithm to find the optimal d, by trying all 2^m possible sets of countermeasures to implement. Our tool called *ADTop* will be published as open-source.

General Optimization Problem. Notice that the generalized optimization problem for selecting countermeasures (considering various assumptions discussed in Sect. 5) can be also solved by applying the approaches from [2,25]. To apply these algorithms under the assumption of *local* countermeasures, we can consider all security controls that have positive effectiveness and their combinations as candidate defense nodes. Furthermore, in case of [2], we will also need to evaluate the resulting set of Pareto-efficient trees, and to select the one that gives the global optimum to the ROSI function.

7   Illustration on a Use Case

We have applied our approach to a use case scenario of a private cloud attack. The target of this scenario is a small/medium size enterprise (SME) with ten employees sharing confidential documents, such as audit reports, studies, and internal documents of customers. To allow continuous remote access to all documents, they are made available on a private cloud accessible via VPN and installed in the SME's IT room. Suppose that stealing these documents will create a damage of 100.000 €.

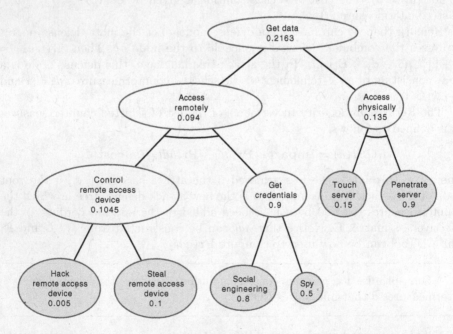

Fig. 1. Initial attack tree with success probabilities for our private cloud attack use case.

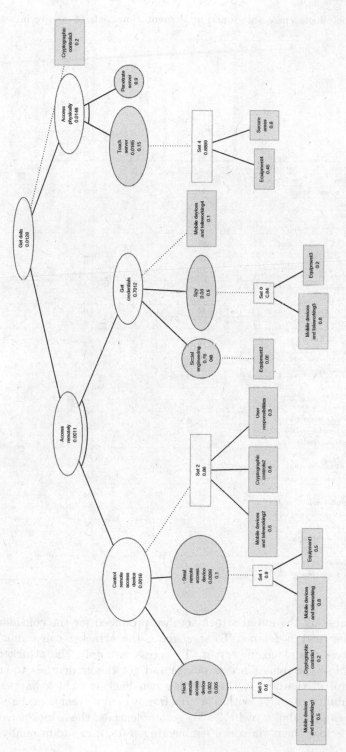

Fig. 2. Optimal attack-defense tree of our use case of the private cloud attack.

Table 1. The effectiveness values and implementation costs of countermeasures.

	Control remote access device	Get credentials	Get data	Social engineering	Penetrate server	Access physically	Access remotely	Hack remote access device	Spy	Steal remote access device	Touch server	Cost in k€
Management direction for inf. security	0.8			0.8					0.2	0.8		3.3
Internal organization			0.3			0.1						0.66
Mobile devices and teleworking	0.5	0.1						0.5	0.8	0.8		0.45
Prior to employment												0.1
During employment												0.2
Termination and change of employment												0.05
Responsibility for assets												0.375
Information classification												0.33
Media handling								0.15				0.4
Business requirements of access control	0.1	0.2						0.05		0.05	0.4	0.96
User access management							0.2			0.2	0.3	2.16
User responsibilities	0.3											
System and application access control	0.7	0.55		0.1				0.7	0.5			1.86
Cryptographic controls	0.6		0.2					0.2				0.5
Secure areas											0.8	0.27
Equipment				0.05					0.2	0.5	0.45	0.53
Operational procedures and responsibilities												3.96
Protection from malware	0.4	0.05		0.1				0.5				0.8
Backup												
Logging and monitoring												0.18
Control of operational software	0.1					.		0.15				0.2
Technical vulnerability management	0.35	0.05						0.3	0.05			0.66
Information systems audit considerations												0.312
Network security management								0.5				0.225
Information transfer												0.156
Sec. requirements of information systems	0.04	0.04		0.05				0.05	0.05			0.4
Sec. in development and support processes								0.05				0.78
Test data												
Inf. security in supplier relationships												0.8
Supplier service delivery management												
Management of incidents												0.6
Information security continuity												1.2
Redundancies												
Compliance with laws and contracts												0.3
Information security reviews												2.4

Figure 1 presents the initial attack tree we produced for the considered use case. It can be read as follows. To steal data, the attacker can remot7ely or physically access the cloud file server. To access remotely, the attacker needs to gain control of the remote access device and get the credentials to connect. To gain control of the device, the attacker can hack it (which happens at a success probability of 0.5 % within a timeframe of one year), or he/she can steal it (success probability of 10 %). To get credentials, the attacker can make the user to disclose them via social engineering (80 %), or, additionally to the hacking, he/she can spy on the privileged user, e.g., by installing a key logger, or, before stealing the remote access device, by spying on the keyboard, e.g., via shoulder surfing (50 %). To access physically, the attacker needs to touch the server(15 %) and to penetrate it, e.g., by plugging a USB stick or accessing the hard disk (90 %). The probabilities were estimated by the customer, for her implementation. The overall success probability of the root node "Get data" is . 21.63 %, computed in the ADTool.

We consider as potential countermeasures the objectives taken from the ISO/IEC 27002 standard (see Table 1). The customer has partially implemented them, and has estimated the security implementation costs to achieve full compliance to these objectives. We have evaluated the effect of these countermeasures on each attack node of the initial attack tree by filling the matrix **E**, which was filled for the eleven attack nodes and the thirty-five countermeasures, i.e., the thirty-five objectives of the ISO/IEC 27002 standard. We identified the objectives without any effect on the attack nodes and removed them, reducing the complexity of the algorithm from 2^{35} to 2^{17} attack-defense trees to consider. The optimal attack-defense tree adt_{opt} found by our implementation is presented in Fig. 2.

Figure 3 shows a screenshot of our ADTop tool that implements the approach described in this paper. The optimal attack-defense tree adt_{opt} found by ADTop has the residual success probability for the attacker reduced to 1.28 % (instead of the initial 21.63 %). The optimization function is computed as **Impact· Probability**(adt_{opt}) + **cost**(selected countermeasures). For the optimal attack-defense tree it is 100,000€ · 0.0128 + 1750 = 3030. The corresponding ROSI is **Impact· (Probability**(at) - **Probability**(adt_{opt})) - **cost**(selected countermeasures) = 100,000€ · (0.2163 - 0.0128) - 1750€ = 18,600€.

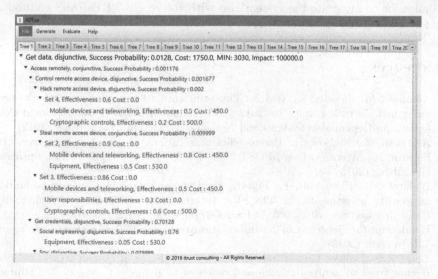

Fig. 3. Screenshot of the ADTop tool.

8   Next Steps and Conclusions

In this position paper we have argued that there is a gap between practical risk assessment methods and academic research. This gap explains why, on the one hand, the practical impact of academic results is somewhat limited, while, on the other hand, practical risk assessment methods do not include state-of-the-art scientific results. Various factors influence this discrepancy. An example is the use of different ontologies, leading to different interpretations of used notions, such as combined defensive mechanisms (meta-defense nodes). Another possible factor is implied by the fact that practical risk assessment methodologies often have a wider scope than specific academic developments, which leads to an interfacing problem between the two.

We argue that an important step forward can be made by bridging this gap through extending practical methods with recent academic results. As an example, we have looked at the extension of the TRICK methodology with recent results on optimal countermeasure selection. In order to do so, we had to agree on a common terminology and had to relate practical design details (like countermeasure catalogues) to academic concepts (like attack-defence trees). In this paper we provided a high-level description of the proposed extension of TRICK and a special algorithm which has been implemented and is being tested in the context of cloud security.

The next steps will focus on improving scalability by designing a better optimization algorithm, and assessing whether the attack-defense-refined risk assessment can be considered more reliable by the risk managers than the established ALE in TRICK Service. Another future extension of this work is to consider attack trees and attack-defense trees automatically generated from some system model [7] as the starting point, instead of manually designed attack trees. This will allow us to integrate the system also with the recent TREsPASS methodology for assisted risk assessment [27].

References

1. Albanese, M., Jajodia, S., Noel, S.: Time-efficient and cost-effective network hardening using attack graphs. In: 2012 42nd Annual IEEE/IFIP International Conference on Dependable Systems and Networks (DSN), pp. 1–12. IEEE (2012)
2. Aslanyan, Z., Nielson, F.: Pareto efficient solutions of attack-defence trees. In: Focardi, R., Myers, A. (eds.) POST 2015. LNCS, vol. 9036, pp. 95–114. Springer, Heidelberg (2015)
3. Bistarelli, S., Fioravanti, F., Peretti, P.: Defense trees for economic evaluation of security investments. In: The First International Conference on Availability, Reliability and Security, 2006 ARES 2006, pp. 8-pp. IEEE (2006)
4. Bundesamt fur Sicherheit in der Informationstechnik: IT-Grundschutz-Catalogues, 13th version (2013)
5. Edge, K.S., Dalton, G.C., Raines, R.A., Mills, R.F., et al.: Using attack and protection trees to analyze threats and defenses to homeland security. In: Military Communications Conference 2006. MILCOM 2006, pp. 1–7. IEEE (2006)

6. European Organization for Safety of Air Navigation: Threats, Pre-controls and post-controls catalogues (2009)
7. Gadyatskaya, O.: How to generate security cameras: towards defence generation for socio-technical systems. In: Mauw, S., et al. (eds.) GraMSec 2015. LNCS, vol. 9390, pp. 50–65. Springer, Heidelberg (2016). doi:10.1007/978-3-319-29968-6_4
8. Gadyatskaya, O., Jhawar, R., Kordy, P., Lounis, K., Mauw, S., Trujillo-Rasua, R.: Attack trees for practical security assessment: ranking of attack scenarios with ADTool 2.0. In: Agha, G., Van Houdt, B. (eds.) QEST 2016. LNCS, vol. 9826, pp. 159–162. Springer, Heidelberg (2016). doi:10.1007/978-3-319-43425-4_10
9. Harpes, C., Adelsbach, A., Zatti, S., Peccia, N.: Quantitative risk assessment with ISAMM on ESA's operations data system. In: Proceedings of TTC (2007)
10. ISO: 27799:2008 Health Informatics - Information security management in health using ISO/IEC 27002 (2008)
11. ISO, IEC: 27005:2011 Information technology Security techniques Information security risk management (2011)
12. ISO, IEC: 27001:2013 Information technology - Security techniques - Information security management systems - Requirements (2013)
13. ISO, IEC: 27002:2013 Information technology Security techniques Code of practice for information security controls (2013)
14. ISO, IEC: TR 27019:2013 Information technology Security techniques Information security management guidelines based on ISO/IEC 27002 for process control systems specific to the energy utility industry (2013)
15. Kordy, B., Kordy, P., Mauw, S., Schweitzer, P.: ADTool: security analysis with attack-defense trees. In: Joshi, K., Siegle, M., Stoelinga, M., D'Argenio, P.R. (eds.) QEST 2013. LNCS, vol. 8054, pp. 173–176. Springer, Heidelberg (2013)
16. Kordy, B., Mauw, S., Radomirović, S., Schweitzer, P.: Attack-defense trees. J. Logic Comput. **24**(1), 55–87 (2014)
17. Mauw, S., Oostdijk, M.: Foundations of attack trees. In: Won, D.H., Kim, S. (eds.) ICISC 2005. LNCS, vol. 3935, pp. 186–198. Springer, Heidelberg (2006)
18. NATO Research and Technology Organisation (RTO): Improving common security risk analysis (2008)
19. NIST: Special Publication 800–53 Revision 4. Security and privacy controls for federal information systems and organizations (2013). http://nvlpubs.nist.gov/nistpubs/SpecialPublications/NIST.SP.800-53r4.pdf
20. NIST: Framework for Improving Critical Infrastructure Cybersecurity (2014)
21. OWASP: CISO AppSec Guide: Criteria for managing application security risks (2013)
22. PCI Security Standards Council: Payment Card Industry Data Security Standards (PCI DSS) (2016). https://www.pcisecuritystandards.org/
23. PWC: The global state of information security survey (2016). http://www.pwc.com/gx/en/issues/cyber-security/information-security-survey.html
24. Refsdal, A., Solhaug, B., Stølen, K.: Cyber-Risk Management. Springer Briefs in Computer Science. Springer International Publishing, Heidelberg (2015)
25. Roy, A., Kim, D.S., Trivedi, K.S.: Scalable optimal countermeasure selection using implicit enumeration on attack countermeasure trees. In: Proceedings of the 42nd Annual IEEE/IFIP International Conference on Dependable Systems and Networks, pp. 299–310. IEEE (2012)
26. Schneier, B.: Attack trees. Dr. Dobb's J. Softw. Tools **24**, 21–29 (1999)
27. TREsPASS: Technology-supported Risk Estimation by Predictive Assessment of Socio-technical Security (2016). http://www.trespass-project.eu/

Enterprise Architecture-Based Risk and Security Modelling and Analysis

Henk Jonkers[✉] and Dick A.C. Quartel

BiZZdesign, P.O. Box 321, 7500 AN Enschede, The Netherlands
{h.jonkers,d.quartel}@bizzdesign.com

Abstract. The growing complexity of organizations and the increasing number of sophisticated cyber attacks asks for a systematic and integral approach to Enterprise Risk and Security Management (ERSM). As enterprise architecture offers the necessary integral perspective, including the business and IT aspects as well as the business motivation, it seems natural to integrate risk and security aspects in the enterprise architecture. In this paper we show how the ArchiMate standard for enterprise architecture modelling can be used to support risk and security modelling and analysis throughout the ERSM cycle, covering both risk assessment and security deployment.

Keywords: Enterprise architecture · Archimate · Risk and security modelling · Risk analysis

1 Introduction

Until quite recently, IT security was the exclusive domain of security specialists. However, due to the fact that the complexity of (networked) organizations and their IT infrastructure is growing, and cyber attacks are getting more sophisticated, traditional approaches to cyber security no longer suffice. In the last couple of years, organizations have started to realize that IT-related risks cannot be seen in isolation, and should be considered as an integral part of Enterprise Risk and Security Management (ERSM). ERSM includes methods and techniques used by organizations to manage all types of risks related to the achievements of their objectives.

It is only natural to place ERSM in the context of Enterprise Architecture (EA), which provides a holistic view on the structure and design of the organization. Therefore, it is not surprising that EA methods such as TOGAF [6] include chapters on risk and security (although the integration of these topics in the overall approach is still open for improvement), and a security framework such as SABSA [5] shows a remarkable similarity to the Zachman framework for EA. And as a corollary, it also makes perfect sense to use the ArchiMate language [8], the standard from The Open Group for enterprise architecture modelling, to model risk and security aspects as an integral part of the architecture.

In this paper, we introduce this risk and security "overlay" of the ArchiMate language (Sect. 2), and link these concepts to the phases of a typical ERSM

© Springer International Publishing AG 2016
B. Kordy et al. (Eds.): GraMSec 2016, LNCS 9987, pp. 94–101, 2016.
DOI: 10.1007/978-3-319-46263-9_6

process (Sect. 3). Subsequently, we show how the resulting models can be used as input for qualitative risk analysis, inspired by the Open FAIR Body of Knowledge [7] (Sect. 4). Using this analysis, the impact of different control measures to mitigate the identified risks can also be assessed. We illustrate this approach with a small example in Sect. 5. Finally, in Sect. 6, we draw some conclusions and give some pointers to other possible applications of enterprise architecture-based risk and security models.

2   Modelling Risk and Security in the ArchiMate Language

The ArchiMate language [8] is the leading open standard for enterprise architecture modelling, aimed at creating integrated models of the organization structure and business processes, supporting software applications and technology, and underlying technical infrastructure, as well as the business motivation and implementation and migration aspects. Although the ArchiMate language does not natively support risk and security modelling, guidelines for using specializations of ArchiMate concepts for this purpose have been published in a white paper from The Open Group [1].

To identify the relevant concepts in the ERSM field, several leading standards and frameworks for risk and security have been studied, including the ISO/IEC 27001 standard on information security management, the Open FAIR Body of Knowledge [7], and the SABSA framework [5], as well as scientific frameworks such as ISSRM [2]. The concepts found in these standards and frameworks show a lot of overlap, and it appears that most of the concepts used in these standards and frameworks can easily be mapped to existing ArchiMate concepts, as summarized in Fig. 1 (the original ArchiMate concepts are shown in brackets). Since ERSM is concerned with risks related to the achievement of business objectives,

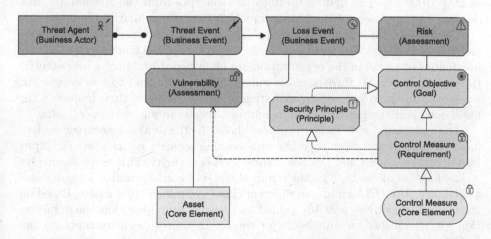

Fig. 1. Risk and security concepts as specializations of ArchiMate concepts

it is not surprising that many of these concepts are from the ArchiMate motivation extension, but also some of the elements from the core language (business, application and technical infrastructure layer) are used:

- Any core element represented in the architecture can be an *asset*, i.e., something of value susceptible to loss that the organization wants to protect. Assets may have *vulnerabilities*, which may make them the target of attack or accidental loss.
- A threat may result in *threat events*, targeting the vulnerabilities of assets, and may have an associated *threat agent*, i.e., an actor or component that (intentionally or unintentionally) causes the threat. Depending on the threat capability and vulnerability, the occurrence of a threat event may or may not lead to a *loss event*, i.e., an actual negative impact caused by the threat.
- *Risk* is a (qualitative or quantitative) assessment of probable loss, in terms of the loss event frequency and the probable loss magnitude.
- Based on the outcome of a risk assessment, we may decide to either accept the risk, or set *control objectives* (i.e., high-level security requirements) to mitigate the risk, leading to requirements for *control measures*. The selection of control measures may be guided by predefined *security principles*. These control measures are realized by any set of core elements, such as business process (e.g., a risk management process), application services (e.g., an authentication service) or nodes (e.g., a firewall).

In the following sections, we will show how these concept can be used for modelling and analysis in the different phases of the ERSM process.

3 The ERSM Process

Figure 2 sketches a typical iterative ERSM process, inspired on standards such as ISO 31000 [3]. The figure also links the concepts from the ArchiMate "risk overlay" to the phases of the process in which they are primarily used.

The left-hand side of this process (phases 1–4) are concerned with risk assessment. Based on monitoring, experience or inspection of the model, potential vulnerabilities of assets in the organization are identified. Combined with potential (internal or external) threats, these vulnerabilities may lead to loss events. An assessment of these loss events, consisting of an indication of their frequency (or likelihood) and the potential loss magnitude, results in an overview of risks.

The right-hand side of the process (phases 5–9) are about security deployment. The identified risks, together with existing security policies, are the input for the control objectives, i.e., the desired level of security. This may also involve a classification of assets, e.g., the required levels of confidentiality, integrity and availability (the "CIA triad") of different classes of information assets. Based on the control objectives, possibly guided by security principles that the organization has established, requirements for control measures (security controls) can be formulated. Ultimately, these control measures are designed and implemented

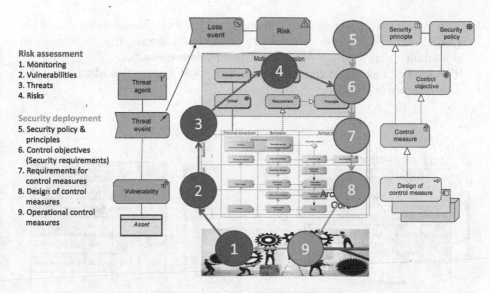

Fig. 2. The ERSM process

within the organization. This leads to a new baseline situation, which forms the starting point of a new iteration of the ERSM process.

In the next two sections we will outline how ArchiMate models can be used in the risk assessment and in the security deployment phases, respectively.

4 Qualitative Risk Analysis

Using the language customization mechanisms as described in the ArchiMate standard [8], risk-related attributes can be assigned to the concepts introduced above. The Factor Analysis of Information Risk (FAIR) taxonomy [7], adopted by The Open Group, provides a good starting point for this. If sufficiently accurate estimates of the input values are available, quantitative risk analysis provides the most reliable basis for risk-based decision making. However, in practice, these estimates are often difficult to obtain. Therefore, FAIR proposes a risk assessment based on qualitative (ordinal) measures, e.g., threat capability ranging from 'very low' to 'very high', and risk ranging from 'low' to 'critical'. Figure 3 shows how these values can be linked to elements in an ArchiMate model, how they are related, and how they can be visualized in 'heat maps':

A. The level of *vulnerability* (Vuln) depends on the *threat capability* (TCap) and the *control strength* (CS). Applying control measures with a high control strength reduces the vulnerability level. In case of multiple threats or multiple control measures, we assume that the maximum threat capability and maximum control strength determine the outcome, although more advanced ways to combine them are also conceivable.

B. The *loss event frequency* (LEF) depends on both the *threat event frequency* (TEF) and the level of *vulnerability*. A higher vulnerability increases the probability that a threat event will trigger a loss event.

C. The level of *risk* is determined by the *loss event frequency* and the *probable loss magnitude* (PLM).

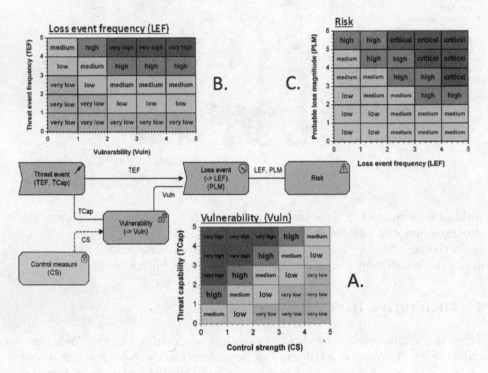

Fig. 3. Summary of qualitative risk analysis

5   Example

The example in Fig. 4 shows a simple application of a vulnerability and risk assessment. The "traffic lights" show the ordinal values of the risk attributes as defined in the FAIR Body of Knowledge and summarized in Sect. 4.

A vulnerability scan of the transmission of payment data from a web shop to an online payment provider has shown that the encryption level of transmitted payment records is low (e.g., due to an outdated version of the used encryption protocol). This is classified as a vulnerability level 'high'. Also, the transmission channel using the public internet is insecure, which is classified as a vulnerability of level 'medium'. These two vulnerabilities enable a man-in-the-middle attack,

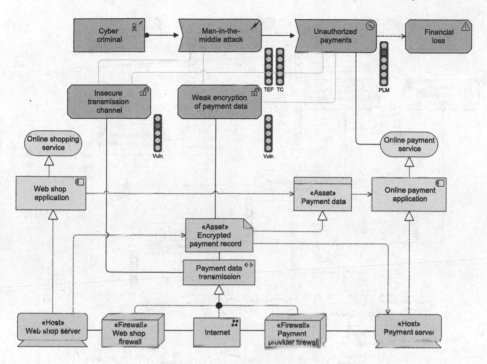

Fig. 4. Risk analysis example

in which a cyber criminal may modify the data to make unauthorized payments, e.g., by changing the bank account number of the receiver. Assuming a cyber criminal with medium skills (medium threat capability) and a medium threat event frequency (e.g., on average one attempted attack per week), according to the loss event frequency matrix shown in Fig. 3, the expected loss event frequency is also medium. Finally, assuming a high probable loss magnitude (potentially, a large sum of money may be lost), the resulting level of risk is high.

It is decided that this risk is unacceptable. Therefore, a control objective is defined to prevent unauthorized access to payment data, together with a security profile specifying the required security parameters for payment data: confidentiality and integrity must be high (it should not be possible for unauthorized persons to view or modify the data), and the required level of availability is medium (payment data does not have to be available 24/7). This is illustrated in Fig. 5. This profile can be translated to specific requirements for control measures. For example, as a preventive control measure that helps to achieve the required levels of confidentiality and integrity, a stronger encryption protocol is needed (which can be realized by, e.g., 256-bit encryption instead of 128-bit encryption), and a secure transmission channel is needed (which can be realized by using a VPN solution). By modifying the parameters, it can be shown what the effect of the different control strengths is on the residual risk. Further reduction of this risk may also require other measures, e.g., measures to limit

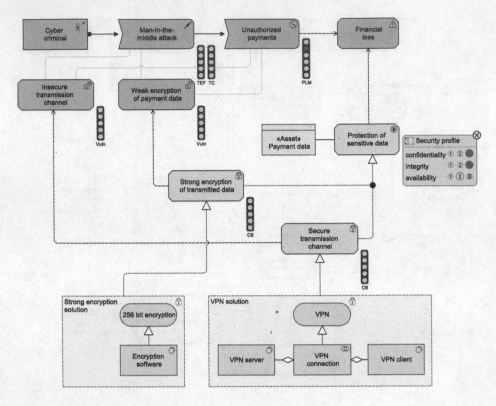

Fig. 5. Risk mitigation example

the probable loss magnitude (e.g., by limiting the maximum amount of money that can be transferred using this system).

6   Conclusions

Because of the increasing complexity of organizations and their IT infrastructure, and the growing capabilities of cyber attackers, traditional information security approaches no longer suffice: it becomes necessary to adopt an integrated approach to Enterprise Risk and Security Management (ERSM).

The ArchiMate modelling language provides the hooks to integrate risk and security aspects in the overall enterprise architecture. By linking risk-related properties to specializations of ArchiMate concepts, risk analysis can be automated with the help of a modeling tool. In this way, it becomes possible to analyze the impact of changes in these values throughout the organization, as well as the effect of potential control measures to mitigate the risks. For example, the business impact of risks caused by vulnerabilities in IT systems or infrastructure can be visualized in a way that optimally supports security decisions made by managers.

The modelling concepts and analysis technique described in this paper have been implemented as a prototype in BiZZdesign's modelling tool suite Enterprise Studio. The approach and tool have been applied in a real-life case study to set up an initial security architecture at a health insurance company. The focus of this case study was on the systematic identification of control objectives and requirements for control measures, and a gap analysis between the baseline and target security architectures. This aspect is underexposed in this paper, but the presented modelling concepts are also very suitable to support this. Another option that has been explored is the import of the results of an automated vulnerability scan (penetration test) of the IT infrastructure into an ArchiMate model, thus making it possible to visualize the found vulnerabilities and their impact throughout the rest of the enterprise architecture [4].

Acknowledgement. Part of the research leading to these results has received funding from the European Union Seventh Framework Programme (FP7/2007–2013) under grant agreement no. 318003 (TRESPASS). This publication reflects only the authors views and the Union is not liable for any use that may be made of the information contained herein.

References

1. Band, I., Engelsman, W., Feltus, C.S., Gónzàlez Paredes, S., Hietala, J., Jonkers, H., Massart, S.: Modeling enterprise risk management and security with the ArchiMate language. White Paper, The Open Group (2015)
2. Dubois, E., Heymans, P., Mayer, N., Matulevičius, R.: A systematic approach to define the domain of information system security risk management. In: Nurcan, S., et al. (eds.) Intentional Perspectives on Information Systems Engineering, pp. 289–306. Springer, Heidelberg (2010)
3. ISO, Itc, UNIDO: ISO 31000 Risk Management: A Practical Guide for SMEs (2015)
4. Jonkers, H., Seghers, B.: Visualizing the business impact of technical cyber risks. In: The Open Group Summit Amsterdam, and as an Open Group Webinar (2014)
5. Sherwood, J., Clark, A., Lynas, D.: Enterprise security architecture. White Paper, SABSA Institute (2009)
6. The Open Group: TOGAF® Version 9.1. Van Haren Publishing, Zaltbommel (2011)
7. The Open Group: Risk taxonomy (O-RT), version 2.0 (2013)
8. The Open Group: ArchiMate® 3.0 Specification. Van Haren Publishing, Zaltbommel (2016)

From A to Z: Developing a Visual Vocabulary for Information Security Threat Visualisation

Eric Li[1,2], Jeroen Barendse[1(✉)], Frederic Brodbeck[1], and Axel Tanner[3]

[1] LUST, The Hague, Netherlands
{eric,jeroen,frederic}@lust.nl
[2] Princeton University, Princeton, NJ, USA
eyli@princeton.edu
[3] IBM Research – Zurich, Rüschlikon, Switzerland
axs@zurich.ibm.com

Abstract. Security visualisation is a very difficult problem due to its inherent need to represent complexity and to be flexible for a wide range of applications. As a result, many current approaches are not particularly effective. This paper presents several novel approaches for visualising information security threats which aim to create a flexible and effective basis for creating semantically rich threat visualisation diagrams. By presenting generalised approaches, these ideas can be applied to a wide variety of situations, as demonstrated in two specific visualisations: one for visualising attack trees, the other for visualising attack graphs. It concludes by discussing future work and introducing a novel exploration of attack models.

Keywords: Visualisation · Security · Model · Attack tree · Attack graph

1 Introduction

Understanding and assessing security threats has always been a key challenge to security practitioners, exacerbated by the emergence of the digital era and the increasingly intricate systems that make up the information security landscape. As a result, developing methods that distill vast amounts of data into consumable visualisations or diagrams that are both engaging and informative remains a critical problem. Many security models[1] can be thought of as a giant machine with tens, or even hundreds of levers and dials that must all be precisely calibrated in order to model each specific security scenario. This is particularly true of the models being developed for use in the TRE$_S$PASS project[2], which aims to

---

[1] A model is defined as "a simplified description, especially a mathematical one, of a system or process, to assist calculations and predictions" (Oxford English Dictionary). When discussing visualisation of said models, it is in regards to making this abstraction visible in some manner.

[2] Technology-supported Risk Estimation by Predictive Assessment of Socio-technical Security http://www.trespass-project.eu.

© Springer International Publishing AG 2016
B. Kordy et al. (Eds.): GraMSec 2016, LNCS 9987, pp. 102–118, 2016.
DOI: 10.1007/978-3-319-46263-9_7

create socio-technical models for security and risk estimation and in which this research was performed. Because of the increasing complexity of security models, it is well worth spending the time to develop visualisations that complement such models.

Visualisation is defined as (i) the formation of mental visual images and (ii) the act or process of interpreting in visual terms or of putting into visible form.[3] As it relates to information security, both senses of the word carry equal weight. Visualisation is used not merely for aesthetics, but also to aid practitioners and end users in forming mental models by providing a visual aid for the data presented in a security model. Because the data presented in models such as Attack Trees [14] tends to be tedious and rather complex, it is important that a visualisation provides an abstraction that reduces the complexity while still maintaining sufficient semantic detail.

Ideally, the visualisation also creates a narrative of the data in a consumable or even actionable manner. It has the power to strongly influence a viewer's perception of a model and therefore requires careful consideration of all the dimensions presented (or not presented). However, many traditional visualisations that have been used to visualise security models tend to fall short of this goal, allowing visualisation of only two, sometimes three, parameters of this multidimensional data. While they do a good job illustrating the relationship between some aspects of their respective security models, attempts to visualise a complete model or any more aspects will not be as simple.

In the next section, we present an overview of current approaches for attack visualisation, explain our approach in Sect. 3, and provide examples of using this approach in Sects. 4 and 5, followed by conclusions and ongoing explorations in Sect. 6.

2   Survey of Current Approaches to Attack Visualisation

In the beginning of the TRE$_S$PASS project, one of the first steps was a survey of state-of-the-art information security risk visualisations [3].

In general, information security visualisations depend very much on the purpose (exploratory versus explanatory), topic (financial risks, environmental risks, computer security etc.), and target audience, with very different levels of abstraction in presenting the vast amount of data typically available for the systems under consideration. Therefore they range from dashboard-like presentations of overall system state for awareness in an operations centre, to tools for investigating very specific technical details, like packet flows across networks, for deep diving into available data. This breadth makes it difficult to survey the complete field. Some summarising reviews can be found in [6,13] and [8], for more general risk visualisations and in [12] for the more technical oriented visualisations in computer security.

Here we focus on a critical review of tools currently used by security practitioners such as Carnegie Mellon's OCTAVE [1] and Siemens' CRAMM [2].

---

[3] As defined by Merriam Webster.

These findings, which serve as background and motivation for a new approach, are summarised here.

Information security visualisations have traditionally been used to display the degree of impact, measure of risk, and value of assets. Tools similar to those mentioned previously use visualisations that map assets to threats and vulnerabilities, and often appear in dashboards. These visualisations cover a wide range of graphic outputs, including visual metaphors to convey certain portions of their security model. However, in most cases, visualisation approaches focus more strongly on functional implementation and interaction than on aesthetics and the narrative defined by the aesthetics. Many of these visualisations typically do not provide a narrative for the motivations of attackers and defenders and consequently require the users to perform their own analysis to draw meaningful conclusions.

Often, visualisations depict the information security attack surface as having only two parameters, requiring researchers to sometimes oversimplify a model to represent it. However, in doing so, crucial interrelationships between actors and elements of the system are not shown, and there is a risk of misrepresenting the data or portraying it in a way that causes the viewer to misinterpret it. As a result, it would be beneficial to explore methodologies of extending existing visualisations to support higher dimensionality, more depth and influencing connections. Visualisations must therefore be flexible, supporting visualisation of individual aspects of the model as well as the model in its entirety. This does not necessarily mean that practitioners should aim to visualise all available data, but to choose the most important aspects to convey the message, visualise this in the most relevant and convincing way, and/or allow viewing from different perspectives.

3 A Parameterised Approach

3.1 Defining a Security Language

Taking language as a metaphor, before writing sentences and paragraphs, one has to first define an alphabet. This alphabet is the domain from which everything else in the language is formed. The richer the words, the more eloquent the sentences that can be written. We are therefore in the process of learning how to 'speak' by developing a language and using the corresponding 'characters' to express the aspects of information security: transforming data into information, information into knowledge, and knowledge into insight. All of these elements themselves can be relatively simple and straightforward. But when combined in an open-ended manner, they can convey complex information.

More concretely, this means to take the whole of a model and break it down into its most elemental components. It is sensible to identify elements or parameters of a model that have an effect on its outcome. This allows the assignment of a certain hierarchy or ranking to the elements, depending on how influential they are, i.e., picking and choosing the particular aspects to focus on and visualise. For example, a typical step in the TRE$_S$PASS Attack Tree, part of its security model, has multiple parameters such as cost, time, probability, and difficulty. These all can affect the outcome of the risk analysis for the model and thus form the basis of the attack tree language.

3.2   A Visual Vocabulary

A visual vocabulary is composed of a set of symbols or graphics that function as building elements for describing larger, more complex visual entities. A strong visual language forms an important basis for representing security models as it provides a suitable mapping from the model's language to the visual vocabulary. This vocabulary should also be extensible, allowing one to highlight and visualise particular parts of interest, including uncertainty. Although uncertainty may seem to go against highlighting important aspects in a security model, both are critical for the understanding and evaluation of information, and should be developed in concert with the remainder of the vocabulary. The core of any visualisation is the selection and development of an effective visual vocabulary and a mapping, or legend, that supports it. Such visual vocabularies are often aided by the principles of *Gestalt* psychology.

Gestalt and Visual Thinking. The overall appearance and qualities, or *Gestalt*, of a visualisation are very important properties. *Gestalt* is a term from psychology defined as the 'unified whole'. Being aware of and implementing the principles of *Gestalt* theory in a visual language renders visualisations stronger and more informed. These theories of visual perception were first developed by a group of German psychologists [10,11] in the 1920s and describe how people tend to organise visual elements into groups. Although there are certain faults with some Gestaltist assumptions [16], it is important to be aware of those principles in order to use and, at other times, also to creatively mis-use them:

Similarity. The principle of similarity states that things sharing visual characteristics such as shape, size, colour, texture, value or orientation will be seen
   as belonging together.
Continuation. The principle of continuity predicts the preference for continuous figures.
Closure. The principle of closure applies when viewers tend to see complete
   figures even if part of the information is missing.
Proximity. The principle of proximity or contiguity states that things which
   are closer together will be seen as belonging together.
Figure and Ground. The terms figure and ground explain how viewers use
   elements of the scene which are similar in appearance and shape and group
   them together as a whole. Similar elements are contrasted with dissimilar
   elements (ground) to give the impression of a whole.
Pre-attentive Variables and Layering. Pre-attentive variables operate mostly at a 'subconscious' level; people recognise trees, tables, and maps, and
   immediately process the underlying data according to the first impressions

gained without any conscious analysis of actual data. Encoding via pre-attentive factors relates to the general graphic design concept of 'layering'. When looking at well-designed graphics of any sort, different classes of information are perceived on the page. Pre-attentive factors like colour cause visuals to perceptually 'pop out,' and any sense of similarity causes them to be seen as connected to one another, as if each were on a transparent layer over the base graphic. This is an extremely effective way of segmenting data, where each layer is simpler than the whole graphic, and the viewer can study each layer in turn, while relationships among the whole are preserved, emphasised, and therefore are brought seamlessly to the analyst's attention.

Pre-attentive variables, combined with certain cultural habits (the colour red indicates stop or dangerous), can already lead to a basic understanding of a visualisation by viewers. Therefore it is important that these pre-attentive variables and habits correspond to rather than contradict the mapping chosen. To be able to fully understand a visual vocabulary developed, every visualisation needs a legend (which is in fact derived from the latin word *legenda*, meaning "the things that need to be read"). A legend or key is often a box in the corner of a map or visualisation, and is essential for understanding a visualisation. Having an effective legend is crucial because it requires the designer to establish clear goals in order to provide a clear mapping from the language of the model to the visual vocabulary. It defines which aspects of the model are represented and how they appear visually. As the legend is actually the most important element on a visualisation, it should be the starting point for each project. This means that the legend need not be a box in the corner, but can be a part of the information presented and as integrated as possible.

3.3   Approaches for Developing Visualisations

Parameterisation of Visual Elements. Every type of visualisation or graph will contain graphic elements whose rendering is modified by variables. A circle, for example, has several variables, such as position, radius, fill and opacity, that affect the final visual outcome. One can think of these variables as knobs that can be adjusted depending on a certain input. This way of thinking allows the creation of nearly unlimited combinations of visual elements.

For example, directed graphs[4] are used heavily in security models because of their ability to represent complex network relationships between entities. These graphs consist of edges and vertices, visually often represented by the two essential elements of a line and a circle. Feedback from early explorations found that it is most visually effective to parameterise at most three variables simultaneously. A line can be defined, for example, by its thickness, colour, and opacity. By applying a mapping from a derived security vocabulary (e.g. difficulty, time, and

---

[4] Here we discuss graphs in the mathematical not the visual, context. So a graph in this context is defined as a network of vertices or nodes connected by (directed or undirected) edges.

probability) to the visual vocabulary, it is possible to begin building the framework for visualisation. Thickness can be mapped to indicate difficulty, colour to indicate time, and opacity to indicate probability. Note that certain properties are more suitable for mapping to certain visual parameters. It does not, for example, make sense to map time to opacity because a lower opacity implies a weak link, whereas time only indicates the length to complete. More examples can be found in [5]. All of these parameters combined lead to a very rich, yet simple visualisation that leaves no visual 'stone' unturned. One can apply a similar approach to visualise information in a node.

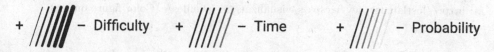

Fig. 1. Legend for attack trees; difficulty is indicated as stroke width; time is indicated as stroke colour; probability as stroke opacity. (Color figure online)

Stacking Visual Elements. There are often cases where the number of parameters that need to be visualised far outnumber the sensible variants to a particular visual element. In other cases, it may not be known what the total number of parameters is (as they may shift depending on user input). An example might be an entity, such as an attacker profile, for which the user has the freedom to pick and choose the parameters to assign. In such cases, it makes sense to develop a language that is extensible. An approach in which visual elements are stacked provides a solution to the problem, as it can be applied to a wide range of parameters.

First, it is important to establish a unified legend, where, for example, line thickness and colour are used to indicate levels of risk. Quite often parameters in security models can be mapped to a scale that ranges from low to high risk. This means that a visual element becomes a generalised module for visualising, allowing for adaptability and re-usability. Attacker profiles are a good example because the number of parameters change depending on the situation. Intel provides a good set of baseline attacker profiles in [5]. But there are cases where perhaps some parameters may not matter. As a result, it is necessary to create a visual system that allows for this. By using a unified legend, as described above, where thickness and colour can represent threat level, it is possible to represent an attacker profile as a set of stacked circles, in which each parameter is one of the circles (Fig. 2). This technique allows extensibility if say, later on, a situation calls for an additional parameter by providing the ability to stack an additional circle. Again, it is important to pay attention to visual hierarchy as parameters that are closer to the outside of a circle are weighted as visually more important. This can be compensated by arranging the parameters in order of importance, or preference, from inside to outside.

108   E. Li et al.

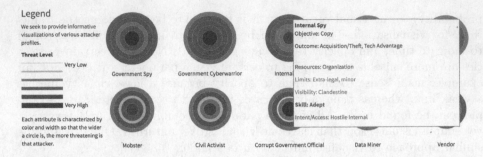

Fig. 2. Legend for stacking circles, and resulting visualisation of attacker profiles. Demo at: http://lustlab.net/dev/trespass/visualizations/profiles/. (Color figure online)

Semantic Zooming. Often, security visualisations will either be too simple in their attempts to abstract a model, or too complex and confusing by trying to show all the data. An approach to solve this issue is the idea of semantic zooming, which applies meaning to different zoom levels. Semantic zooming is an approach that displays different, semantically relevant, information at different levels of zoom on a visualisation. More abstract representations can be used at a macro view, whereas all the complex intricacies can be shown in a micro view. This is similar to the appearance of additional details when zooming into an online map.

This approach complements stacking visual elements quite well as it allows elements to be seen only when such detail is required. For example, when viewing attacker profiles from a macro view, it is only important to show the total perceived threat that an attacker has. As a result, the attacker can be represented by a single circle whose radius is the sum of the stacked circles. Only when zoomed into a more detailed view, where a viewer might want to inspect how parameters differ between attackers, does the visualisation reveal the individual stacked elements. By displaying this detail only when necessary, it is possible to create visualisations that can be relatively simple without any loss of information while still allowing the data to be understood in detail through taking a closer look.

Multiple Views. Every visualisation foregrounds certain aspects of the data it is representing, while backgrounding other aspects. As not all viewers are interested in the same aspects, multiple views offer a good way to cater to this. In addition to multiple views on the same data, it is also possible that a certain model can be analysed by various tools. Each of these tools provides its own outcomes and therefore needs its own visualisation, because each view tackles a certain aspect of the security model. Viewed as a whole, often in a dashboard-like view, they paint the entire picture of the visualisation.

Contextual Awareness and Highlighting. A key aspect in security visualisation is the ability to highlight key points of vulnerability. This tends to be much more effective than just textual output, as it also gives viewers the ability to contextualise potential points of interest in the model. Oftentimes there are analytical tools that can provide insights such as the weakest or cheapest path in an attack tree. It may also make sense to highlight certain connections based on user interaction. It is also important to consider how and when certain elements will be highlighted when developing a vocabulary. Depending on what needs to be highlighted, certain approaches may be better than others. Contextual awareness also allows a fine-grained representation of information without overwhelming the viewer.

As mentioned, visualising uncertainty is often just as important as the data itself, as it allows viewers to understand the accuracy or the fuzziness of certain factors of the model. Existing work in visualising uncertainty can be found in [7, 9]. Possibilities to visualise uncertainty include transformations such as blurring visual entities, or introducing a way of displaying multiple possible predictions of a model, similarly to how a user might choose to highlight a certain path or visual element.

4   Application to Attack Trees in the TREsPASS Project

Attack trees, as developed and used in the TREsPASS consortium, are a tool to capture all possible attacks to reach a specific goal, as described in the root node. To build such an attack tree, experts typically gather and, starting from the goal node, try to enumerate possible ways of attack to reach this goal. Each subnode can be iteratively refined as far as it seems fit. Individual intermediate nodes can thereby be either conjunctive or disjunctive. A conjunctive node requires that all of its children be fulfilled in order to proceed up the tree, whereas disjunctive nodes only require one child to be satisfied in order to proceed. A complete path of actions consists of any number of leaf and intermediate nodes leading to the root node. Within the project, each leaf node is considered to contain four parameters: difficulty, minimum cost, probability of success, and minimum time required to complete. Depending on the effort put into the creation, these attack trees can be very complex, comprising hundreds or thousands of nodes, especially when they are generated programmatically from an underlying model as it is the aim of the project. When attempting to visualise these trees, such visualisations quickly become complex and unreadable.

In their traditional form, attack trees present a wide variety of important and relevant information, but are not easily visualised, oftentimes shown as an arrangement of text in a directed graph. From a visualisation perspective, attack trees have several flaws; the tree structure gets very wide rapidly, repeating lots of elements to eventually become effectively unreadable even in a medium allowing arbitrary zooming. Also, because attack trees consist of conjunctive and disjunctive nodes, it needs to become visually clear that in the case of conjunctive nodes, all steps need to be fulfilled in order to proceed. We can counteract the complexity by improving the way the tree is laid out and labelled, as well as by testing alternative layouts that result in more compact trees, while maintaining

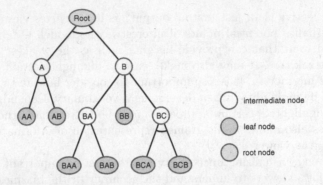

Fig. 3. Example of a typical structure of an attack tree.

readability. Next to that, exploring interactivity allowing the user to zoom and pan, and to collapse sub-trees at any level, makes it easier to concentrate only on certain parts of the tree (Fig. 3).

The key components and their respective properties are the following:

Node	Edge
Type of node (leaf, intermediate, root)	Parent/Child nodes
Conjunctive or disjunctive node (if intermediate or root node)	
Label	
Minimum cost to complete node	
Probability of success	
Difficulty	
Minimum time required to complete	

A visual language can be developed based on this. It is important to keep in mind that attack trees can vary greatly in size, as their construction is largely dependent on the scenario and environment that they are trying to model. As a result, the language should be scaleable to any size tree. In initial explorations, feedback revealed that when representing the graph with directed graphs, edges carry much more weight and information visually. Subsequently, most of the parameters were mapped to the edge leading to parent nodes. This allows focus to be placed much more on the path rather than each individual step. The resulting legend was developed by parameterising the visual properties of a line (Fig. 1) and creating a mapping to the attack tree vocabulary (Fig. 4).

4.1   Multiple Views

Visualising an attack tree in a tree structure may do a good job for displaying how the nodes are connected, but it does a poor job for examining frequency.

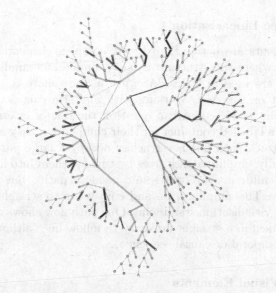

Fig. 4. Attack tree visualised in radial form where each node corresponds to an attack step. The root node, i.e., the goal, is placed in the center. The edges are coloured according to three parameters: difficulty is indicated as stroke width, time as stroke colour, and probability as stroke opacity. (Color figure online)

Therefore, it makes sense to split this into two visualisations: an attack tree visualisation structured as a tree, as well as a tree map visualisation that focuses just on the relative frequency of each node (Fig. 5). The frequency of the node determines the size of each box, while the colour depicts the relative difficulty of each node. A hover over each box in the tree map shows its label and highlights the nodes in the attack tree, allowing a user to understand the visualisation. Together, they paint a more complete picture. We consider the tree map as part visualisation and part legend.

Fig. 5. Tree map visualisation that shows frequency of an attack step. (Color figure online)

4.2 Attack Tree Linearisation

More simple elements are better than fewer, complex elements. A tree works well in situations where the structure is fairly simple and small. However, the attack trees that are used in TRE$_S$PASS are already more complex than can comfortably be fit on a screen. Working with and studying attack trees from a visualisation point of view, one can question the role of intermediate nodes. Other than being a labelled container for their child nodes, they are not actually steps along the attack path but nevertheless occupy a large part of the attack tree. We can visually simplify attack trees by turning them into linear sequences of their required children. This will result in more paths, but each path will be easier to follow. The simplification and conversion to straight paths benefit readability from a visualisation standpoint. One path now shows a user the steps that need to be taken in a straight and easy to follow line (although it does not usually imply a temporal or causal sequence).

4.3 Stacking Visual Elements

Another legend was also developed in the case that additional parameters to each step may be needed. By mapping different visual elements (thickness and colour to threat level) of a line to a scale of threat, it is possible to modularise this element and stack it to any number of parameters.

Visually, this becomes just as effective as the original legend because a step in which all parameters have a high perceived threat level will stand out much more strongly than a step with a low perceived threat level. When combined to form a path, as in Fig. 6, this legend is very informative on which steps and connections are areas of vulnerability.

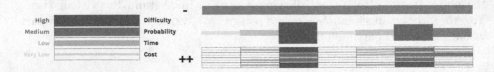

Fig. 6. Tree map plus radial attack tree visualisation, where the size in the tree map shows the frequency of an attack step and the colour indicates difficulty. Both views are linked in highlighting when selecting elements. (Color figure online)

4.4 Semantic Zooming

When applying the stackable legend developed in Sect. 4.3 to massive attack trees, the overall visual effect of the visualisation becomes confusing and harder to read. Viewers are not as able to follow paths as easily as before. However, semantic zooming can be applied in multiple ways. Because the stacked lines are only necessary at a very detailed level, it is perfectly fine to show the average threat level at a macro view with other paths or the entire tree (Fig. 6).

Only when zooming in to view specific paths will the individual stacked lines be revealed to the viewer. This eliminates the original complexity at a macro level while still allowing specificity at a micro level.

This can be combined with a rearrangement of the linearised attack trees to present the paths in a more understandable manner. By using a radial view for the linearised attack trees at a macro view and transitioning to a table, in which information about the total path can be displayed alongside each path upon user interaction, it can be possible to sort and analyse paths in a way that might otherwise be unwieldy at the macro level (Fig. 7). A viewer can then zoom in even closer to see an individual path and its stacked line components, as well as any intermediate labels that might not have been shown before.

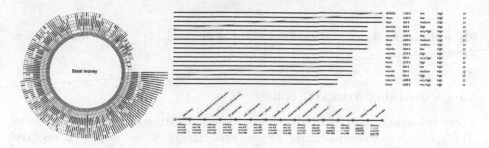

Fig. 7. Semantic zooming applied to linearised attack paths. From left to right: an attack tree in radial and straight path form, and the same paths as a table allowing detailed inspection

5   Application to Attack Graphs

Attack graphs are a common tool used by security researchers to organise information on all possible attack paths within a certain space. Although they generally are adapted for custom use, the general idea is the same: there is a directed graph with a starting point and an end point (the goal), as well as nodes that function as attack steps or entities (Fig. 8) (as used in [4]). The edges are possible paths from one entity to another. These nodes and edges carry with them several parameters, such as probability, cost, and incident count.

For the remainder of this section, however, the attack graph referred to is the one defined by Verizon in their annual Data Breach Investigations Report (DBIR)[5] [15] as a test case to see whether we could also apply the visualisation principles laid forward in the preceeding chapters to graphs other than attack trees. For 2016, there are seven action groups, with multiple sub-actions, as well as three attribute groups, again with multiple sub-attributes. Within the graph itself, actions lead to other actions or to compromised attributes. Compromised attributes will lead either to the end of the breach or to another action by the attacker.

[5] Data link: https://github.com/vz-risk/VERISAG/tree/v2/static.

Fig. 8. Example attack graph as used in the online tool *2016 DBIR Attack Surface Analysis* [4].

5.1   Visualising Attack Graphs

There are several goals that the visualisation of the attack graph aims to achieve: (i) displaying and differentiating actions and attributes, (ii) displaying relative threat of nodes and edges, (iii) displaying paths, and (iv) displaying a comparison between different versions of the graph (either through mitigations or comparison with previous years' data). The principal flaw of traditional attack graph visualisations is that they attempt to visualise all nodes and connections at once. In cases such as the DBIR, this grows very complex and as a result, it becomes hard to perform even simple tasks, such as determining the relative importance of a node or discovering which nodes are connected. In fact, the version presented at [4] mostly serves to illustrate how complex the attack space is.

As the principal structure is a directed graph, it is possible to immediately identify the nodes and edges as elements of its vocabulary. Digging deeper, the following characteristics make up these elements:

Node	Edge
Type of node (action, attribute, start, end)	Type of edge (edge to attribute, action, or end)
Relative frequency of node (occurrences within incidents)	Relative frequency of edge (occurrences within incidents)
Sub-type of node (one of 5 actions or one of 3 attributes)	

The visual language begins with the same traditional elements of the directed graph: nodes and directed edges. Traditionally, these graphs are visually composed of circles that represent the nodes, and paths with arrows indicating direction.

To begin building a visual vocabulary, each of the elements is parameterised. Assigning radius and fill colour of the circle to represent frequency of incidents creates an aesthetically informative visualisation of the node. Textual treatment and visual treatment can then be applied to each circle to indicate the type of node. Rather than by drawing an arrow as a path, direction on edges can be shown by decreasing the stroke width of a path. This width can also be parameterised, as well as the opacity of the edge, to show how frequently that edge occurred within the incident space, resulting in the legend in Fig. 9. From this legend the visualisation[6] was developed based on the outlined approach.

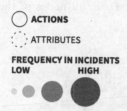

Fig. 9. Legend for attack graph visualisation. (Color figure online)

Arc Diagram. Arc diagrams were chosen because of their ability to clearly display multivariate data, even in complex situations. Building on the work of Wattenberg [17], these graphs provide an easy way to visualise connections by placing the nodes on the same line (Fig. 10). They also allow easy comparison of nodes and edges, and give viewers a more linear ability to visualise this data. The two-sided nature of this graph allows one to visualise edges going to attributes on one side, and edges going to actions on the other, providing a clear visual delineation between the two types of edges. When a comparison of different versions of the graph should be shown, all edges can be moved to one side and two versions of the graph can be shown simultaneously.

Semantic Zooming. Although visualisation of all 119 nodes in one view provides a good overview, this level of complexity is very hard to follow and unactionable. To improve this, semantic zooming can be applied. The initial view now becomes the eight macro categories of actions and attributes. This presents a good overall sense of how certain attacks paths might be structured, as well as the relative frequency of certain action or attribute categories within the attack space. If viewers want to learn more about the composition and frequency of each attack or attribute, they can click on it for a view containing all subgroups (Fig. 11a). We choose to contain these sub-nodes within the overall node to visually show the hierarchy of nodes. At this level, viewers can see the count of each node and determine which specific sub-node presents a greater threat. This also applies to comparing graphs (Fig. 11b).

---

[6] The interactive version of the DBIR Attack Graph can be found at http://lustlab. net/dev/vzw/index.html.

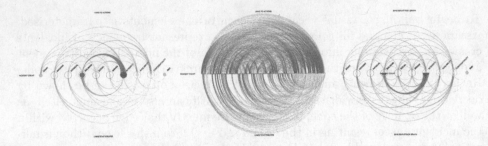

Fig. 10. Different visualisation views afforded by using an arc diagram. From left to right: macro-view of 2016 DBIR Data, all nodes of the 2016 DBIR, and comparison between 2015 and 2016 DBIR.

Fig. 11. From left to right: (a) Micro-view of 2016 DBIR data with highlighting, (b) Micro-view of 2015/2016 DBIR data with highlighting, (c) Highlighting a specific node and its target edges in the 2016 DBIR. (Color figure online)

Contextual Awareness and Highlighting. A level of interactivity is also built into the visualisation that allows the viewer to highlight certain aspects of the graph depending on the current zoom level. At the macro-view, hovering over a node reveals the incident count of that node within the attack space, and other child nodes (Fig. 11c). This allows viewers to pay attention to the context in which they are highlighting the node. All other nodes are greyed out for clarity, which further aids focusing on the relevant information. In the micro-view, when a user hovers over a node in the graph, the incident count at the bottom of the circle updates to also display information specific to the currently highlighted node (Fig. 11a). Visual feedback is also given by turning the highlighted node white to match the text colour. This contextual highlighting provides information to the user only when requested and presents the wealth of information in the graph in a non-overwhelming manner.

6   Conclusion and Future Work

Security-related visualisations have always been a particular challenge, due in part to the complex, multi-dimensional, and wide ranging nature of the field. Not only do models tend to be complex in order to capture the intricacies of a particular scenario, but their usage varies from highly specialised technical people requiring very specific visualisation to security researchers attempting to create layman infographics that aim to visualise an overall model, resulting in visualisations that are not quite effective because they try to do too much or too little. As designers begin to move into the era with interactive visualisations, we should consider how all of these differing aspects will play a role in our approach to security visualisation.

This paper presented an approach for developing effective visualisations of complex security models. By first outlining goals for visualisation and then breaking down a model and defining a visual vocabulary, it is possible to create a framework that allows nearly endless flexibility. From here, a visualisation can be built by applying some combination of the approaches outlined. These approaches offer ways of thinking about visualising complex data that are general enough to be applied to almost any type of visualisation situation, including those not related to security. We use attack trees from the TRE$_S$PASS project as well as attack graphs from Verizon's DBIR report as case studies to see how this parameterised approach can be used to create effective visualisations.

Early versions of this visualisation approach have been presented to two security practitioner feedback panels and the advisory board of the TRE$_S$PASS project. Based on their feedback, visualisations were adjusted and developed further. For instance, the stacking of visual elements is a direct result of feedback given, as many practitioners had a hard time combining more than two parameters in one visualisation.

Currently we are implementing features in the code so that external data can be loaded easily and can be applied to situations in which the input data is dynamic and changing frequently. In that way, this visualisation approach can cover an extensive range of media, from print to interactive and to time-based. We can also then see if the concepts and vocabulary developed are applicable to non-security-related sectors. The final code and designs will be made available as open source so that third parties can use and extend this work.

Acknowledgements. The research leading to these results has received funding from the European Union's Seventh Framework Programme (FP7/2007–2013) under grant agreement ICT-318003 (TRE$_S$PASS). This publication reflects only the authors' views, and the European Union is not liable for any use that may be made of the information contained herein.

References

1. Alberts, C.J., Dorofee, A.: Managing Information Security Risks: The Octave Approach. Addison-Wesley Longman Publishing Co., Inc., Boston (2002)
2. Barber, B., Davey, J.: The use of the CCTA risk analysis and management methodology (CRAMM) in health information systems. Medinfo **92**, 1589–1593 (1992)
3. Barendse, J., Bleikertz, S., Brodbeck, F., Coles-Kemp, L., Heath, C., Hall, P., Kordy, B., Tanner, A.: TRE$_S$PASS Deliverable 4.1.1: initial requirements for visualisation processes and tools . Internal deliverable of the TRE$_S$PASS project. (2013)
4. Bassett, G.: Verizon Enterprise Solutions: DBIR Attack Graph Analysis, June 2015. http://dbir-attack-graph.infos.ec/

5. Bertin, J.: Sémiologie Graphique. Gauthier-Villars, Paris (1967)
6. Eppler, M.J., Aeschimann, M.: Envisioning risk: a systematic framework for risk visualization in risk management and communication (2008). http://www.knowledge-communication.org/pdf/envisioning-risk.pdf
7. Harris, R.L.: Information Graphics: A Comprehensive Illustrated Reference. Oxford University Press Inc., New York (1999)
8. Husdal, J.: Can it be really that dangerous? Issues in visualization of risk and vulnerability (2001). http://www.husdal.com/2001/10/31/can-it-really-be-that-dangerous-issues-in-visualization-of-risk-and-vulnerability
9. Kirk, A.: References for visualising uncertainty. http://www.visualisingdata.com/2015/02/references-visualising-uncertainty/
10. Koffka, K.: Principles of Gestalt Psychology. Harcourt, Brace and Company, New York (1935)
11. Koffka, K.: Perception: an introduction to the Gestalt-theorie. Psychol. Bull. **19**(10), 531–585 (1922)
12. Marty, R.: Applied Security Visualization, 1st edn. Addison-Wesley Professional, Boston (2008)
13. Roth, F., Eidgenössische Technische Hochschule (Zürich), Crisis and Risk Network, Schweiz. Bundesamt für Bevölkerungsschutz, Suisse. Office Fédéral de la Protection de la Population: Visualizing risk: the use of graphical elements in risk analysis and communications. 3RG report, Eidgenössische Technische Hochschule Zürich, Center for Security Studies CSS (2012). http://e-collection.library.ethz.ch/view/eth:6286
14. Schneier, B.: Attack trees: modeling security threats. Dr. Dobb's J. Softw. Tools **24**(12), 21–29 (1999). https://www.schneier.com/cryptography/archives/1999/12/attack_trees.html
15. Verizon Enterprise Solutions: 2016 Data Breach Investigations Report. Technical report, Verizon (2016). http://www.verizonenterprise.com/verizon-insights-lab/dbir/
16. Ware, C.: Information Visualization: Perception for Design. Morgan Kaufmann Publishers Inc., San Francisco (2000)
17. Wattenberg, M.: Arc diagrams: visualizing structure in strings. In: IEEE Symposium on Information Visualization, 2002, pp. 110–116. IEEE. (2002)

Quantitative Attack Tree Analysis: Stochastic Bounds and Numerical Analysis

Nihal Pekergin[1([⊠])], Sovanna Tan[1], and Jean-Michel Fourneau[2]

[1] LACL, Université Paris-Est Créteil, Créteil, France
nihal.pekergin@u-pec.fr

[2] DAVID, Université de Versailles-St-Quentin, Versailles, France

Abstract. This paper presents an efficient numerical analysis of the time dependence of the attacker's success in an attack tree. The leaves of the attack tree associated with the basic attack steps are annotated with finite discrete probability distributions. By a bottom-up approach, the output distributions of the gates, and finally the output distribution at the root of the attack tree is computed. The algorithmic complexities of the gate functions depend on the number of bins of the input distributions. Since the number of bins may increase rapidly due to the successive applications of the gate function, we aim to control the sizes of the input distributions. By using the stochastic ordering and the stochastic monotonicity, we analyze the underlying attack tree by constructing the reduced-size upper and lower distributions. Thus at the root of the attack tree, we compute the bounding distributions of the time when the system would be compromised. The main advantage of this approach is the possibility to have a tradeoff between the accuracy of the bounds and the algorithmic complexity. For a given time t, we can compute the bounds on the probability for the attacker's success at time t. The time-dependent behavior of attacks is important to have insights on the security of the system and to develop effective countermeasures.

Keywords: Attack tree · Discrete probability distribution · Stochastic bounds

1 Introduction

The notion of Attack Trees (AT) has been popular since the seminal work of Schneier [10] to model security threats and/or risk analysis. Attack Trees are non-state-space models which illustrate graphically complex scenarios. They have been essentially proposed for computer security analysis but they can also be applied for quantitative risk analysis. The probability at the root represents the risk that the modeled system is subjected to. The goal is to enhance the threads of the system, and to illustrate the impact of countermeasures of basic events on

This work was partially supported by grant ANR MARMOTE (ANR-12-MONU-0019).

the global behavior of the system in order to define the priorities for reinforcing the subsystems of the model.

A rich survey on the attack and defense models based on Directed Acyclic Graph (DAG) models is given in [5]. The first models are static and have been proposed in early 90s to adopt tree-based models of reliability engineering to security engineering. These models are called threat trees, vulnerability trees, attack trees, and they describe complex attack scenarios from the basic events and logical connectors AND and OR. The probabilities of the occurrences of basic events are the input parameters, and the output parameter is the probability of the event at the root of the underlying tree.

It is important to evaluate the success probabilities of an attack as a function of time. As it has been the case for the reliability analysis, dynamical models have been proposed by including the time to success of an attack, and also new connectors taking into account temporal dependencies.

Attack Trees and Fault Trees have apparent similarities. In a Fault Tree, component faults lead to high level faults. However in an Attack Tree the actions are taken intentionally by attackers, while faults happen in a fault tree. In a Fault Tree, the model is in general a consequence of the system architecture, the safety may be reinforced by considering more reliable components and/or by including the spares. In an Attack Tree a countermeasure implementation as a consequence of the risk analysis may introduce other possible attacks. Thus the system architecture may need to be revised.

Safety and security have developed as two distinct disciplines for many years. Thus analysis methods and dedicated tools have been developed in each field by separate communities. However they are closely related and share many commonalities. These observations have been made by many authors. For instance, see [9] and the references therein for the cross-fertilization between safety and security engineering.

The methods that have been proposed to analyze quantitatively Dynamical Fault Trees (DFT) have been extended for the quantitative analyze of attack trees [2,8]. This paper is the extension of the methodology proposed by some of us [3] for efficient numerical analysis of DFT to the quantitative risk analysis for systems specified by an attack tree. In [2], the quantitative analysis of attack trees has been performed approximatively to overcome the algorithmic complexities. The basic attack distributions are supposed to be acyclic phase type continuous distributions, and the output distributions of gates obtained from the gate functions are approximated as acyclic phase type distributions. In this work we apply a similar idea but our technique is radically different. First we consider discrete probability distributions, and control the increase of the size of the distributions by replacing them with reduced-size bounding distributions in the sense of the $\leq_{st}$ order. Intuitively speaking, if two distributions are ordered in the sense of this order: $d_1 \leq_{st} d_2$, then it means that the cumulative probability distribution of d_1 is always greater or equal to the cumulative probability distribution of d_2. In other terms, d_2 takes larger values than d_1. The monotonicity properties of gates (AND, OR, SEQ) have been proved in [3]: if the input distributions are replaced by bounding distributions the output provides a bounding distributions. Thus we derive *lower*, and *upper* bounding, reduced-size distributions not the exact one. The system is then evaluated approximatively, however the results provide bounds. By increasing the number of bins, it is possible to improve the tightness of the bounds.

The paper is organized as follows. Section 2 is devoted to the Attack trees and their evaluation. We first present Attack trees, and then explain our methodology based on the use of bounding discrete distributions. In Sect. 3, two case studies from the literature are analyzed with the proposed methodology.

2   Attack Trees and Evaluation

2.1   Attack Trees

The Attack Tree (AT) is a tree composed of basis attacks (BA) and logical gates. The leaves are the basic attack steps. The complex attack scenarios are specified by combining basic attacks with logical gates. Thus an AT may lead to decompose complex scenarios into easier, understandable, quantifiable actions. In the formalism of [10] the logical gates consist of conjunction (AND) and disjunction (OR) gates. Sequential conjunction (SEQ) gates have been included in [4].

Input distributions associated with the leaves represent the time that the input will become $True$ (the time at which the underlying BA would be successful). Similarly, the output distribution of a gate corresponds to the time at which the subsystem having this gate as root would be compromised. Therefore the output distribution of the root of an Attack Tree denotes the time when the attack for the whole system would be successful.

We refer to [4] for the general semantics of the attack trees. The temporal semantics of the gates is given in the following. Let X_1 and X_2 be the random variables representing the times that the corresponding input becomes $True$. The inputs are assumed to be mutually independent. The cumulative probability distribution for a random variable X will be denoted by F_X:

$$F_X(t) = \Pr(X \le t).$$

- $AND(X_1, X_2)$: The output of this gate is the random variable O with $O = \max(X_1, X_2)$. The output becomes $True$ at a time when both inputs become $True$. The cumulative distribution of the output is computed by the product of the cumulative distributions of X_1 and X_2.

$$F_O(t) = F_{X_1}(t) \times F_{X_2}(t).$$

- $OR(X_1, X_2)$: The output of this gate is $O = \min(X_1, X_2)$. The output becomes $True$ at a time when one of the inputs become $True$. The reliability (survival) distribution of the output is computed by the product of the reliability distributions of X_1 and X_2.

$$1 - F_O(t) = (1 - F_{X_1}(t)) \times (1 - F_{X_2}(t)).$$

– $SEQ(X_1, X_2)$: The output of this gate is the random variable $O = X_1 + X_2$. For the output becomes *True*, first the first input must become *True*, then the second input becomes *True*. Thus, the output distribution is computed by the convolution of the input distributions.

2.2   Analysis with Discrete Random Variables

The proposed methodology is based on the use of the discrete probability distributions. In the sequel we call them equally histograms. Using histograms would allow us to perform efficient numerical analysis to compute the bounding output distributions of the gates. The histograms associated with the leaves (times for basic attacks) may be general distributions which are not restricted to special types. If the underlying distributions are continuous random variables, the related histograms can be constructed by discretization. Moreover, empirical distributions can be constructed by means of measures and statistical methods, and then they can be inputs of the AT model. We assume that the input distributions are mutually independent. As it is usual in DFT analysis, the temporal analysis is limited by a Mission Time (MT). Thus it means that the analysis is done from time 0 to MT.

The output histogram of a gate is computed from its input histograms, depending upon its type.

– OR gate:

$$\Pr(O = a) = \Pr(X_1 = a) \times \Pr(X_2 > a) + \Pr(X_2 = a) \times \Pr(X_1 > a) + \Pr(X_1 = a) \times \Pr(X_2 = a).$$

– AND gate:

$$\Pr(O = a) = \Pr(X_1 = a) \times \Pr(X_2 < a) + \Pr(X_2 = a) \times \Pr(X_1 < a) + \Pr(X_1 = a) \times \Pr(X_2 = a).$$

– SEQ gate:

$$\Pr(O = a) = \sum_k \Pr(X_1 = k) \times \Pr(X_2 = a - k).$$

The sizes of the output distributions depend on the gates. Let l_1 and l_2 be the respective sizes of input distributions. The maximum size of the output distribution are as follows:

– AND and OR gates: $l_1 + l_2 - 1$. For AND (resp. OR) the minimal (resp. maximal) value among the values of the two histograms is not included in the output histogram.
– SEQ gate: $l_1 \times l_2$.

The computation of the output depends on the sizes of the input distributions. Let $l = \max(l_1, l_2)$. For AND and OR gates, if the input distributions are ordered on the values of the histograms, the output can be computed by an algorithm based on the fusion of the sorted lists ($\Theta(l)$). Note that the input distributions can be sorted within complexity $\Theta(l \times \log l)$. The convolution can be computed by a naive algorithm with complexity $\Theta(l_1 \times l_2)$, and with complexity $\Theta(l \times \log l)$ through Discrete Fast Fourier transformation.

Due to its tree structure, the output distribution of an Attack Tree can be evaluated by a bottom-up approach. However the sizes of histograms (the number of bins) can increase rapidly due to the successive applications of the operations associated with the gates. Especially, the convolution operation associated with SEQ gates may increase the number of bins multiplicatively with the sizes of the input histograms. The time complexity increases with the increase of the histogram sizes.

2.3 Bounding Distributions

The ability to control the sizes of the histograms during the analysis of an AT is of great importance for the point of view of the algorithmic complexity. In [3], in the context of DFT analysis, we have shown that the reduced-size bounding output distributions can be derived by considering bounding input distributions. These bounding distributions are in the sense of the stochastic strong order, $\leq_{st}$. This stochastic order is associated with increasing functions, thus if two random variables are ordered in the $\leq_{st}$ order then their increasing functionals are also ordered:

$$X \leq_{st} Y \quad \Leftrightarrow \quad \mathrm{E}[f(X)] \leq \mathrm{E}[f(Y)] \tag{1}$$

for all increasing function f, when the expectations E exist. For instance, if $X \leq_{st} Y$, then $\mathrm{E}[X] \leq \mathrm{E}[Y]$. More informations can be found in [3,6].

We now explain the methodology of the construction of reduced-size bounding distributions in the context of Attack Tree analysis. Let X be a discrete random variable taking values in a finite set $\mathcal{S}_X$ of size n_X. The probability mass function (histogram) can be given by two vectors $\mathcal{V}_X$ and $\mathcal{P}_X$. The ith entry $\mathcal{V}$, $\mathcal{V}_X[i]$ denotes the ith value that X can take where $\mathcal{P}_X[i]$ denotes the corresponding probability. Without loss of generality, we assume that the vectors are increasingly ordered with respect to the entries of the vector $\mathcal{V}$. In order to reduce the algorithmic complexity to derive the output distribution, only the values for which $\mathcal{P}_X[i] > 0$, $1 \leq i \leq n_X$ are specified in the histograms. Notice that in the context of an AT analysis, the values are the dates at which the basic attacks would be successful if the histograms are related to the leaves, otherwise they are the dates at which the attacks represented by the related subtrees would end up.

Example 1. Let X be the time at which a basic attack (a leaf in the AT) would be successful. Let $\mathcal{S}_X = \{2, 5, 8.5, 10, 15\}$, then $n_X = 5$. The histogram of X can be specified by the following two vectors.

$\mathcal{V}_X$	2	5	8.5	10	15
$\mathcal{P}_X$	0.25	0.2	0.3	0.15	0.1

Therefore the input X turns to *True* at dates $\mathcal{S}_X = \{2, 5, 8.5, 10, 15\}$ with the corresponding probabilities given in the probability vector $\mathcal{P}_X$.

Let U be an upper bounding distribution of X: $X \leq_{st} U$. Then the following inequalities must be satisfied between X and U:

$$i \in \{n_X, n_X - 1, \cdots, 2\}, \quad \sum_{\{k|k \geq i\}} \mathcal{P}_X[k] \leq \sum_{\{j|\mathcal{V}_U[j] \geq \mathcal{V}_X[i]\}} \mathcal{P}_U[j]. \qquad (2)$$

It is assumed that the indexes of vectors start from 1. For the above inequalities index 1 is not considered since the sum of the probabilities must be equal to 1.

Example 2. Let U take values $\mathcal{S}_U = \{4, 10, 15\}$, then $n_X = 3$. It can be checked that the inequalities (2) are satisfied between X and U.

$\mathcal{V}_U$ || 4	10	15
$\mathcal{P}_U$ || 0.25	0.65	0.1

It can be seen from Fig. 1 that the cumulative probability distribution (cdf) of U is always smaller or equal to the cdf of X. (It is more probable to take larger values for U than for X).

The probability that the attack related to the random variable X would be successful at time t can be computed as follows:

$$\sum_{\{i|\mathcal{V}_X[i] \leq t\}} \mathcal{P}_X[i]. \qquad (3)$$

As a direct consequence of Eqs. (2), and (3), we have the following inequalities on the success probabilities of the attacks associated with X and U:

Proposition 1. *If $X \leq_{st} U$, then for every time t, the success probability before or at time t for the attack whose time is associated with X is greater or equal to the success probability before or at time t for the attack associated with U.*

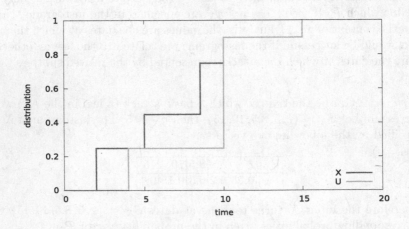

Fig. 1. X and U cumulative distributions.

It has been proved in [3] that the AND, OR, SEQ gates are monotone. Hence if the inputs of these gates are replaced by bounding distributions, then the output distribution computed by the bounding input distributions would provide a bounding distribution on the output distribution computed by the input distributions. This monotonicity follows from the fact that the functions associated with these gates are non decreasing. By using this monotonicity property, it is possible to construct the reduced-size distributions. Obviously, dealing with the reduced-size distributions decreases the underlying algorithmic complexity. In the following example, we illustrate this concept by considering an AND gate.

Example 3. We consider an AND gate with inputs X and Y. The output of this gate is computed first with inputs X and Y and then by taking bounding distributions on X such that $L \leq_{st} X \leq_{st} U$. In the following tables, distributions Y, L are given while X and U are the same as given before.

$\mathcal{V}_Y$	5	7	9	13	20
$\mathcal{P}_Y$	0.3	0.1	0.25	0.15	0.2

$\mathcal{V}_L$	2	7	10
$\mathcal{P}_L$	0.45	0.45	0.1

In the following tables we illustrate the output distributions of respectively $O = AND(X, Y)$, $O_U = AND(U, Y)$, and $O_L = AND(L, Y)$.

$\mathcal{V}_O$	5	7	8.5	9	10	13	15	20
$\mathcal{P}_O$	0.135	0.045	0.120	0.1875	0.095	0.135	0.08	0.2

$\mathcal{V}_{O_L}$	5	7	9	10	13	20
$\mathcal{P}_{O_L}$	0.135	0.225	0.225	0.065	0.150	0.2

$\mathcal{V}_{O_U}$	5	7	9	10	13	15	20
$\mathcal{P}_{O_U}$	0.075	0.025	0.0625	0.4225	0.135	0.08	0.2

In the following figure, we give the cumulative probability distributions of these output distributions. It can be seen that the curve of $AND(L, Y)$ is always above or equal than the other two curves, while the curve $AND(U, Y)$ is always below or equal than the other two curves. Therefore the output distributions are also ordered as follows:

$$AND(L, Y) \leq_{st} AND(X, Y) \leq_{st} AND(U, Y).$$

It follows from Proposition 1, that for a given time t, a lower bound on the attack success probability is obtained from $AND(U, Y)$ while an upper bound on the attack success probability is from $AND(L, Y)$. For instance, at $t = 7$, the success probability for $AND(U, Y)$ is 0.1, for $AND(X, Y)$ is 0.18 and for $AND(L, Y)$ is 0.36.

Using bounding distributions is twofold: first reduced-size bounding distributions can be constructed in order to decrease the algorithmic complexity. On the other hand, it is extremely difficult to estimate temporal behaviors of basic attacks. Deriving bounds in the context of the uncertainty is useful.

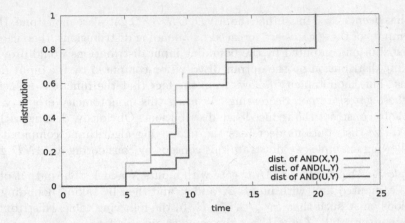

Fig. 2. The distribution of $AND(X, Y)$ and the bounding distributions.

In this paper, the algorithms to construct reduced-size bounding discrete distributions are not presented. Different algorithms to build reduced-size bounding algorithms are given in [1,3]. Intuitively, some bins of the original histogram are deleted, and the related probabilities are moved by taking care of the inequalities associated with the $\leq_{st}$ order.

2.4   Algorithm

In this subsection, we summarize our methodology as an algorithm. It is assumed that the basic attacks which are represented as leaves in the corresponding AT occur mutually independently. Let $\mathcal{D} = d_1, \cdots, d_l$ be the set of distributions for the leaves. Distribution d_i is a finite discrete distribution and it represents the time of success of the ith attack. Let $\mathcal{G} = g_1, \cdots, g_p$ be the set of gates of the AT. An attack tree A is specified by the gates $\mathcal{G}$, the tree structure defining the dependence relations. The distributions associated to the leaves $\mathcal{D}$ are the second set of input parameters.

The output parameter is the distribution at the root of an AT tree, and it is computed by a bottom-up approach. For a gate g, if its input distributions are the distributions at the leaves or are the output of other gates already computed, its output distribution is computed depending on its type. Thus the gates $\mathcal{G}$ are considered in a topological order depending on the structure of the attack tree A.

The algorithmic complexity of the output distribution depends on the sizes (numbers of bins) of the input distributions. We aim to control the sizes of the distributions by constructing the reduced-size bounding distributions. In the previous section, this approach has been explained through some toy examples. The computed results are approximative but they provide bounds. The main idea here is to have a tradeoff between the tightness of the bounds and the algorithmic complexity. Obviously, if the size of distribution is greater, the results would be more accurate. The algorithms to construct bounding distributions are

---

Algorithm 1. Computing bounds on the output distribution of an AT.

Input: AT: A
    input distributions for the leaves: $\mathcal{D}$
    max number of bins of a distribution: $n \in \mathbb{N}$
Output: Output distribution at the root of A.
 1: Label the gates using the topological order from the bottom-up.
 2: **for** all gates g in the ascending order of the labels **do**
 3:    Evaluate the output distribution of gate g
 4:    If the size of the output distribution is larger than n, reduce its size to n.
 5: **end for**

---

given in [1,3]. There are different algorithms with different algorithmic complexities so providing different accuracies. One of them provides optimal bounding distributions with respect to increasing positive rewards (for instance expectation) but with a high complexity ($\Theta(N^2 n)$), where N is the size of the original distribution and n is the size of the bounding distribution). There is a greedy algorithm with complexity ($\Theta(N \log N)$) which can provide sometimes optimal distributions. Algorithms with linear complexities have also been given. Thus depending on the required accuracy and the acceptable complexity, the user can choose one of them.

We now illustrate the usefulness of bounds in the case when the quantitative analysis is done to check some constraints. Let d be the output distribution of the considered attack tree A, and d_L (resp. d_U) be the lower (resp. upper) bounding distributions:

$$d_L \leq_{st} d \leq_{st} d_U.$$

For a given time t, let p be the probability that the system is not compromised at time t, that can be computed from distribution d by applying Eq. (3). Similarly, p_L and p_U are computed through distributions d_L and d_U. It follows from Proposition 1 that

$$p_U \leq p \leq p_L.$$

It is worth emphasizing that if we are only interested in checking if some probabilistic constraints are satisfied or not, the bounds may be sufficient regardless of their accuracy. For instance, if the quantitative evaluation of the attack tree is done to check if the probability that the underlying system is not comprised at time t is less or equal to *threshold* or not, then we can conclude

- if $p_L \leq threshold$, then the constraint is satisfied
- if $p_U > threshold$, then the constraint is not satisfied
- otherwise, the bounds must be refined (the number of bins must be increased)

In this work, as it has been usually the case, we consider that the input distributions of leaves are mutually independent. However this proposed approach can be extended to the case when some basic attacks are dependent by using conditional probabilities. In [3], it has been shown that the construction of bounding distributions in the sense of the $\leq_{st}$ order is compatible with the conditioning to consider the dependent distributions.

3   Case Studies

In this section, we consider two case studies from the literature. The first one is *Steal Exam* which models a student to steal the forthcoming exam [2]. The attack tree given in Fig. 3 is composed of three kinds of attacks represented by three subtrees: social interaction (subsystem S_1), hacking (subsystem S_2), and steal hard copy (subsystem S_3). In subsystems S_1 and S_3, the attack necessitates sequential events, thus the root for these subsystems is a *SEQ* gate. For instance in S_3, first one must *locate office*, then *get access* to it, and finally *find print outs*. For subsystem S_2, either the *mailbox* or *repository* may be hacked, thus the root is a *OR* gate. Success times of basic attacks to success have been inspired from [2] and taken as truncated Erlang, and Exponential distributions. Once the number of bins is fixed, the distributions for the leaves are discretized by taking equal intervals to derive two discrete distributions *inf* and *sup*. For distribution *inf*, within each interval, the value is taken as the smallest value of the interval (left limit) with the probability which is the sum of the probability of the interval. Similarly, for distribution *sup*, the value is taken as the greatest value of the interval (right limit) with the probability of the interval. The missing probabilities for distribution *inf* is added to the smallest time, while it is added to the *MT* for distribution *sup*.

In Fig. 4, we give the bounding cumulative probability distributions of the time to success (in hours) of the AT given in Fig. 3. For this example, the number of bins is limited to 50 (resp. 200) for the leaves and the output distributions of the intermediate gates reach 200 (resp. 1400) bins (for subsystem S_3). Notice that it results from the $\leq_{st}$ order that the distribution *inf* with 50 bins is less or equal in the sense of the $\leq_{st}$ order than all other distributions (its cumulative probability distribution is always greater than that of the others). The distribution *sup* with 50 bins is less or equal in the sense of the $\geq_{st}$ order than all

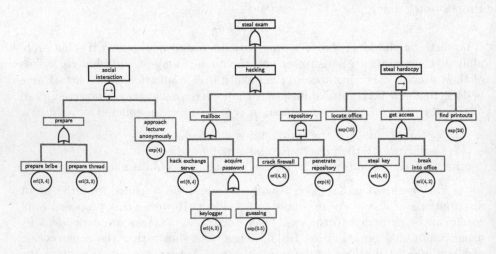

Fig. 3. Attack tree *Steal Exam*.

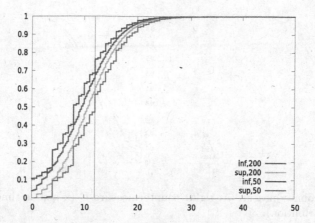

Fig. 4. The output distribution for *Steal Exam* (x axis is time in hours, y axis is the distribution).

other distributions (its cumulative probability distribution is always greater or equal than that of the others). It can be seen that the bounds computed with 200 bins for the distributions of leaves are tighter (they are placed between the distributions with 50 bins). The bounds on the probability that the system is comprised before or equal time t (let say p) can be deduced from these bounding cumulative distributions. For instance, for $t = 12$, from bounding distributions with 50 bins, it can be deduced that $0.57 \leq p \leq 0.76$, while from bounding distributions with 200 bins, it can be deduced that $0.606 \leq p \leq 0.677$. Obviously, the more bins, the tighter the bounds. The quantitative analysis of attack trees are very useful to highlight the impact of the potential countermeasures that can be taken to reinforce the security of the system. We study the impacts of two countermeasures. Countermeasure A consists in preventing the *steal hardcopy*, thus to delete subsystem S_3, while countermeasure B consists in forbidding to send the exams via emails thus to delete subsystem *mailbox*. In Fig. 5, we illustrate the impact of these countermeasures where the size of the basic attack distributions is limited to 200 bins. It can be seen that both countermeasures have effect on the output distribution but the impact of countermeasure B is more important. For instance, the probability that an attack would be successful within time interval 12 h ($0 \leq t \leq 12$) with the original system is in the interval $[0.606 - 0.677]$ and the system with countermeasure A is in $[0.522 - 0.593]$ while it is $[0.45 - 0.523]$ with countermeasure B. Notice that since the ATs for the countermeasures and the original model are not the same, the output distributions are not comparable in the sense of the $\leq_{st}$ order (the distributions of the original model and the countermeasures may cross).

The second example comes from [2,7,11]. The authors modeled the Stuxnet attack against Iranian nuclear enrichment infrastructure which took place in 2010. Stuxnet is a computer worm with a malicious payload designed to compromise the Programmable Logic Controllers (PLCs) and the Siemens

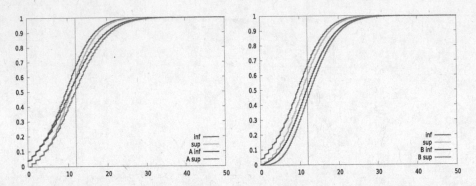

Fig. 5. Left: countermeasure A; right: countermeasure B (x axis is time in hours, y axis is the distribution).

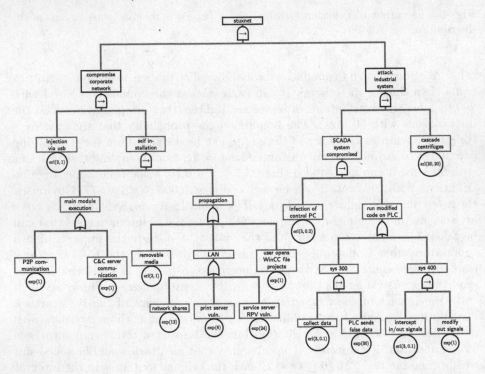

Fig. 6. Attack tree *Stuxnet*.

Supervisory Control And Data Acquisition (SCADA) which controls the centrifuges. The malware damaged the latter while deceiving the control room with false previously collected values. The AT in Fig. 6 consists of two subtrees linked by a *SEQ* gate. The left subtree represents the worm propagation and the right one is the payload attack. Since the target network is not connected to the Internet, the attack begins with an infection through a USB device corresponding to the leftmost BA. As in [2], we studied the impact of the different BAs. For each BA, one at a time, we divided its expected execution time by two and computed the *inf* and *sup* distributions at the AT root and compared it to the original ones. The BA takes place sooner in the modified model.

The following table gives the four most influential BAs to the attack execution time. As in [2], we found that 'collect data', 'infection of control PC' and 'intercept in/out signals' have the greatest impact. Therefore the countermeasures should address them as a priority.

	After 20 days	After 40 days	After 60 days
Original	[0.015, 0.139]	[0.422, 0684]	[0.866, 0.948]
Collect data	[0.055, 0.259]	[0.682, 0.857]	[0.967, 0.989]
Infection of control PC	[0.060, 0261]	[0.678, 0.844]	[0.962, 0.986]
Intercept in/out signals	[0.044, 0.246]	[0.662, 0.850]	[0.964, 0.989]
Injection via USB	[0.022, 0.155]	[0.463, 0.704]	[0.883, 0.953]

In Fig. 7 we compare the original distribution and the distribution when the "collect data" rate is double. The BA are discretized with 25 bins. The distribution bounds at the Attack Tree root have around 1200 bins. With these parameters, the computation takes a few seconds hence the distributions do not need to be compressed during the computation process. Notice that curves for different rate parameters may cross due to the bounding discretization of input distributions. In Fig. 8 we show the results obtained when we increase the BA size to 50 bins. The distribution bounds at the Attack Tree root have around 2500 bins. On our server the computation time is multiplied by a factor 6. When comparing Figs. 7 and 8, we observe the impact of discretization on the accuracy

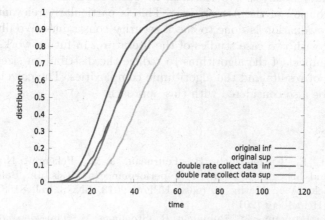

Fig. 7. The output distribution for *Stuxnet* and the distribution when "collect data" has double rate with bin size 25 (x axis is time in day).

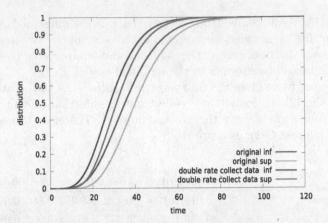

Fig. 8. The original output distribution for *Stuxnet* and the distribution when "collect data" has double rate with bin size 50.

of results. The more precise the discretization is, the more accurate the results are. In both case studies, the reduction of the number of bins has been applied for the leave distributions while the continuous distributions are discretized. The number of bins of output distributions of gates are not reduced due to the rapid overall computation times.

4   Conclusion

The proposed methodology is based on the stochastic comparison to derive bounding distributions on the time to success of Attack Trees. Due to the stochastic monotonicity properties of the *AND*, *OR*, *SEQ* gates, the upper and lower discrete, reduced-size, bounding distributions can be derived. Although an approximative analysis is performed, we build the stochastic bounds on the output distribution of the Attack Tree. This is particularly relevant when the quantitative evaluation is done to check security constraints. We illustrate our methodology with two case studies of the literature. In future work, we aim to analyze the impact of the algorithms to reduce the distribution sizes ([1,3]) on the accuracy of results and the algorithmic complexities. The correlated distributions can be also considered with this approach.

References

1. Aït-Salaht, F., Castel-Taleb, H., Fourneau, J.-M., Pekergin, N.: Stochastic bounds and histograms for network performance analysis. In: Balsamo, M.S., Knottenbelt, W.J., Marin, A. (eds.) EPEW 2013. LNCS, vol. 8168, pp. 13–27. Springer, Heidelberg (2013)
2. Arnold, F., Hermanns, H., Pulungan, R., Stoelinga, M.: Time-dependent analysis of attacks. In: Abadi, M., Kremer, S. (eds.) POST 2014 (ETAPS 2014). LNCS, vol. 8414, pp. 285–305. Springer, Heidelberg (2014)

3. Fourneau, J.M., Pekergin, N.: A numerical analysis of dynamic fault trees based on stochastic bounds. In: Campos, J., Haverkort, B.R. (eds.) QEST 2015. LNCS, vol. 9259, pp. 176–191. Springer, Heidelberg (2015)

4. Jhawar, R., Kordy, B., Mauw, S., Radomirović, S., Trujillo-Rasua, R.: Attack trees with sequential conjunction. In: Federrath, H., Gollmann, D., Chakravarthy, S.R. (eds.) SEC 2015. IFIP AICT, vol. 455, pp. 339–353. Springer, Heidelberg (2015). doi:10.1007/978-3-319-18467-8_23

5. Kordy, B., Piètre-Cambacédès, L., Schweitzer, P.: DAG-based attack and defense modeling: don't miss the forest for the attack trees. Comput. Sci. Rev. **13–14**, 1–38 (2014)

6. Muller, A., Stoyan, D.: Comparison Methods for Stochastic Models and Risks. Wiley, New York (2002)

7. Nielsen, J.R.: Evaluating information assurance control effectiveness on an air force supervisory control and data. Master's thesis, Air Force Institute of Technology (2011)

8. Piètre-Cambacédès, L., Bouissou, M.: Attack and defense modeling with BDMP. In: Kotenko, I., Skormin, V. (eds.) MMM-ACNS 2010. LNCS, vol. 6258, pp. 86–101. Springer, Heidelberg (2010)

9. Piètre-Cambacédès, L., Bouissou, M.: Cross-fertilization between safety and security engineering. Rel. Eng. Sys. Safety **110**, 110–126 (2013)

10. Schneier, B.: Attack trees: modeling security threats. Dr. Dobbs J. Softw. Tools **24**(12), 21–29 (1999)

11. Siwar Kriaa, L.P., Bouissou, M.: Modeling the stuxnet attack with BDMP: towards more formal risk assessments. In: 2012 7th International Conference on Risks and Security of Internet and Systems (CRiSIS), Cork, Ireland, 10–12 October 2012, pp. 1–8. IEEE (2012)

Survivability Analysis of a Computer System Under an Advanced Persistent Threat Attack

Ricardo J. Rodríguez[1]([✉]), Xiaolin Chang[2], Xiaodan Li[3], and Kishor S. Trivedi[3]

[1] Department of Computer Science and Systems Engineering,
University of Zaragoza, Zaragoza, Spain
rjrodriguez@unizar.es
[2] Department of Information Security,
Beijing Jiaotong University, Beijing, People's Republic of China
xlchang@bjtu.edu.cn
[3] Department of Electrical and Computer Engineering,
Duke University, Durham, USA
xiaodan.li@duke.edu, ktrivedi@duke.edu

Abstract. Computer systems are potentially targeted by cybercriminals by means of specially crafted malicious software called Advanced Persistent Threats (APTs). As a consequence, any security attribute of the computer system may be compromised: disruption of service (availability), unauthorized data modification (integrity), or exfiltration of sensitive data (confidentiality). An APT starts with the exploitation of software vulnerability within the system. Thus, vulnerability mitigation strategies must be designed and deployed in a timely manner to reduce the window of exposure of vulnerable systems. In this paper, we evaluate the survivability of a computer system under an APT attack using a Markov model. Generation and solution of the Markov model are facilitated by means of a high-level formalism based on stochastic Petri nets. Survivability metrics are defined to quantify security attributes of the system from the public announcement of a software vulnerability and during the system recovery. The proposed model and metrics not only enable us to quantitatively assess the system survivability in terms of security attributes but also provide insights on the cost/revenue trade-offs of investment efforts in system recovery such as vulnerability mitigation strategies. Sensitivity analysis through numerical experiments is carried out to study the impact of key parameters on system secure survivability.

Keywords: APT · Cyberattacks · Markov chains · Stochastic reward nets · Security metrics · Survivability · Transient analysis

1 Introduction

The number of incidents related to cyberattacks is increasing rapidly, according to numerous reports [1–3]. These cyberattacks have a cost of downtime and

© Springer International Publishing AG 2016
B. Kordy et al. (Eds.): GraMSec 2016, LNCS 9987, pp. 134–149, 2016.
DOI: 10.1007/978-3-319-46263-9_9

cleaning up of compromised systems, besides loss of customer confidence and of other possible long-term consequences due to loss and theft of information. This situation becomes specially critical when cybercriminals attempt to attack infrastructures that provide essential services to the society, such as financial services, power distribution, or water treatment plants [4]. In these systems, an intentional malfunction causing a discontinuity of service may lead to fatalities or injuries. Unfortunately, the number and sophistication of cyberattacks targeting these systems demonstrate an increasing trend [5,6].

Malicious software (*malware*) are pieces of software specially crafted by cyber-criminals to achieve their malicious goals [7]. There exist different types of malware depending on their behavior, such as viruses, worms, botnets, or keyloggers, among others [8]. When malware are designed to target a specific system, they are known as Advanced Persistent Threats (APTs) [9]. The term "advanced" means the target requires a sophisticated attack, since attackers make a previous reconnaissance of the target to know in advance as much as possible about the system to compromise. The term "persistent" means the goal of the threat is to maintain a presence on the targeted system for long-term control and data collection (which are later exfiltrated).

One of the first APTs was *Operation Aurora*, publicized by Google in 2010. Presumably coming from China and with an extremely wide-scale range, it targeted companies of different domains, such as Yahoo, Google, Symantec, Northrop Grumman, Morgan Stanley, and Dow Chemicals [10]. Another well-known APT is the *Stuxnet* attack, also discovered in 2010. This *cyber weapon*, attributed to the US and Israel, was specially designed to exploit Siemens PLCs in SCADA networks affecting Iranian nuclear facilities [9,11]. APTs discovered in the wild from 2007 to 2013 with political intent, such as GhostNet, Flame, or DarkSeoul, among others, are summarized in [12].

An APT comprises of different stages. In *entry point* and *exploitation* stages, the APT gains access to a targeted system by means of zero-day vulnerabilities (i.e., a software flaw that is unknown to the vendor) or vulnerabilities already known but not yet patched. For instance, Stuxnet used four different zero-day vulnerabilities. After gaining access, the APT tries various methods to make itself persistent into the system (*infection* stage) and starts looking for data of interest to be stolen or modified (*lateral movement* stage). Once sensitive data are obtained, the APT will modify or send those data out of the organization's network boundaries (denoted *exfiltration* stage), thus compromising data integrity and confidentiality. In addition, the various actions undertaken during the attack may crash the system and then reduce system availability.

Assessing the impact of APTs on a system is important to characterize the system against these unexpected and intentional failures and to evaluate mitigation techniques that may be applicable. In this regard, survivability refers to a system's ability to withstand malicious attacks and support the system's mission even when parts of the system are damaged [13]. This paper, in particular, defines the *system secure survivability* as a transient measure of the ability of the system to provide pre-specified service with a certain security assurance during the system recovery from a vulnerability.

In this paper, we assess the survivability of a computer system targeted by an APT. A security model is developed to capture both the behavior of the system's response to a security attack and the actions performed by an attacker to cause such an attack. We make a simplifying assumption that all relevant event times are exponentially distributed and thus the model is a homogeneous continuous time Markov chain (CTMC). Note that a number of techniques are available to relax this assumption if needed [14]. We leave the relaxation of this assumption for future work. The generation and solution of the proposed Markov model is automated using a variant of stochastic Petri nets called Stochastic Reward Nets (SRNs). SRNs have been successfully used in the analysis of several domains [15–19], and can easily represent common characteristics of computer systems such as concurrency, synchronization, conditional branches, looping, and sequencing. We furthermore define four survivability metrics (see Sect. 3.1 for details) that account for: (i) system recovery, (ii) system availability, and (iii) data confidentiality and/or integrity loss; after the public announcement of a software vulnerability and during vulnerability mitigation strategy is being deployed.

Related Work. Research has been conducted on survivability modeling and analysis in various fields and from different perspectives [20–24]. Regarding survivability metrics, little research has proposed quantitative evaluation metrics in terms of survivability. Quantitative measures were proposed in [25] to analyze the survivability of a resilient database system against intrusions, modeled with CTMC. This work was later extended to semi-Markov processes in [26]. Similarly, a general approach for survivability quantification of networked systems using SRNs was given in [27]. Survivability assessment of the Saudi Arabia crude-oil pipeline network, modeled with Generalized Stochastic Petri nets, was performed in [28]. All these works only analyze availability under unexpected events.

However, to the best of our knowledge, no work exists that proposes a quantitative assessment of the system secure survivability. The developed model in this paper not only considers the response of the system to a security attack, but also the actions performed by an attacker to cause such an attack and consider the transient behavior of the system in face of an attack. The proposed model and metrics let us investigate system security attributes (namely, confidentiality, integrity, and availability [29]) during the transient period which starts after a vulnerability discovery and through all the stages of an APT, until the vulnerability is fully removed from the system. This paper shows that the results not only enable us to quantitatively assess the system survivability in terms of security attributes but also provide insights on the cost/benefit trade-offs of the investment in system recovery efforts such as vulnerability mitigation strategies.

This paper is organized as follows. Background on Petri nets and Stochastic Reward nets is provided in Sect. 2. The system description and the model considered in this paper are presented in Sect. 3. Then, Sect. 4 deals with numerical results and discussion. Finally, Sect. 5 concludes the paper and outlines future work.

2   Previous Concepts

A Petri net [30] (PN) is a 4–tuple $\mathcal{N} = \langle P, T, \mathbf{Pre}, \mathbf{Post} \rangle$, where P and T are disjoint non-empty sets of *places* and *transitions*, and $\mathbf{Pre}$ ($\mathbf{Post}$) are the pre–(post–)incidence non-negative integer matrices of size $|P| \times |T|$. The *pre-* and *post-set* of a node $v \in P \cup T$ are respectively defined as $^\bullet v = \{u \in P \cup T | (u, v) \in F\}$ and $v^\bullet = \{u \in P \cup T | (v, u) \in F\}$, where $F \subseteq (P \times T) \cup (T \times P)$ is the set of directed arcs.

Graphically, a PN is a bipartite directed graph having two disjoint types of nodes: *places*, drawn as circles; and *transitions*, drawn as bars. A directed arc that connects a place (transition) to a transition (place) is called an input (output) arc of the transition. An arc never connects the same type of nodes. A positive integer inscribed next to an arc specifies the multiplicity associated with the arc. Places that connect to a transition by input arcs are named *input places* of the transition. Similarly, places that are connected to a transition by output arcs are named *output places* of the transition. Each place may contain zero or more tokens, depicted by an integer (or black dots) within the circle representing the place. The number of tokens of a place denotes the *marking* of the place. A *Petri net system*, or *marked Petri net* $S = \langle \mathcal{N}, \mathbf{m_0} \rangle$, is a Petri net $\mathcal{N}$ with an *initial marking* $\mathbf{m_0} \in \mathbb{Z}_{\geq 0}^{|P|}$.

A transition $t \in T$ is *enabled* when each of its input places has, at least, as many tokens as the multiplicity of the corresponding input arc. An enabled transition t can *fire* triggering, upon firing, two actions: first, a number of tokens equal to the multiplicity of the corresponding input arc is removed from each of its input places; and second, a number of tokens equal to the multiplicity of the corresponding output arc is deposited in each of its output places. Thus, the firing of a transition may yield a new marking of the Petri net, named *reached marking*. The *reachability set* is defined as the set of all markings reachable through any possible sequence of transitions, starting from the initial marking $\mathbf{m_0}$.

Stochastic Petri nets are Petri nets where each transition has an exponentially distributed firing time. Generalized Stochastic Petri nets [31] (GSPN) allow transitions that fires in zero time, named as *immediate transitions* and represented by thin black bars. The transitions that follow any distribution firing time are named *timed transitions* and represented by unfilled rectangles. Immediate transitions have always priority over timed transitions to fire. Similarly, immediate transitions with the same input places may have defined a probability to calculate the one that fires when they compete for firing. GSPN also includes inhibitor arcs.

A Stochastic Reward Net (SRN) [32] is a GSPN augmented with reward functions. In SRN, an enabling function (also called a *guard*) defines the enabling function of a transition as a marking-dependent function. In addition, both arc multiplicities and firing rates are allowed to be marking-dependent. SRN allows us to compute measures of interests by defining reward rates at the net level.

3   System Description and Model

To analyze and quantify security attributes of a system, we consider not only the system defender's response to a security attack, but also the actions taken by an attacker to cause the attack. This requires that the security model incorporates the behavior of both elements. In the following, we first describe the system considered in this paper, and then we present a Stochastic Reward Net model for survivability analysis of this system.

3.1   System Description

Figure 1 depicts a flowchart detailing the actions of both the attacker and the defender, and the system changes due to these actions during the recovery from a vulnerability.

Let us consider a system in which an attacker has some interest in accessing into it. The attacker has acquired a previous knowledge of the system, but at the beginning there are no known vulnerabilities the attacker can take advantage of. We consider the attacker is not skillful enough to find unknown vulnerabilities. When a vulnerability is fully disclosed, the system is in the vulnerable state. Meanwhile, the defender starts the patch implementation and the attacker starts the exploit implementation. In the following, we use patch and vulnerability mitigation strategy interchangeably. The shaded part with dotted line in Fig. 1 describes the system state transitions and the shaded rectangles in it represents the system states.

There are eight system states: *System is vulnerable, Patching, Fixing, Failing, Infected, Lmoved, Exfiltrated, Crashing*. The transitions to any of the left four states are triggered by attacker actions. Upon finishing the exploitation software, the attacker starts a sequence of actions to destroy the system security at least in terms of confidentiality, integrity, and availability. These actions include infecting the system, keeping itself persistent in the system, searching sensitive data and making these data benefit them. The last two actions are repeated, forming a loop. Each attacker action may crash the system, denoted by *Crashing* state. In addition, the software bugs, such as Mandelbugs [33], may lead to system failure, denoted by *Failing* state. If the system crashes or fails, it must be fixed immediately. During the fixing, both the defender and the attacker can do nothing to the system. We assume that as long as patch is ready, it must be deployed into the system immediately. Thus, for each attack action, *System fails* and *Patch ready* must be checked in the first place. Since each attacker action may crash the system, for each attack action, we will check whether the attacker action succeeds before a system state transition occurs.

We assume that when the system completes this fixing, there is no APT code in the system but the vulnerability still exists. Such constrain may be relaxed. We leave the modeling of the relaxed system for future work. The system is unavailable before system failure or crash is fixed. Upon completing recovery, the system enters into good state, denoting that the vulnerability does not exist in the system. But when the system fails or crashes, the strategy can only be

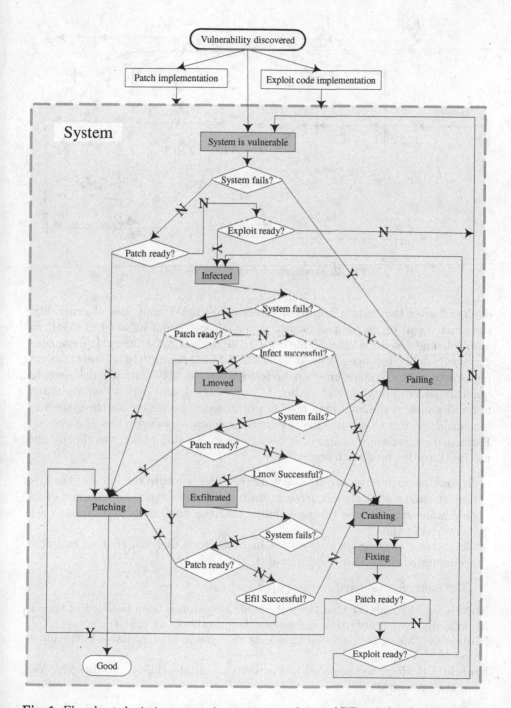

Fig. 1. Flowchart depicting events in a system under an APT attack, after a vulnerability is announced and during the mitigation strategy implementation.

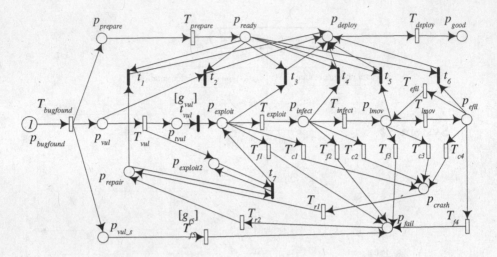

Fig. 2. Stochastic Reward Net model.

deployed after the system failure or crash is fixed. Without loss of generality, the deployment process is assumed to never fail and is not affected by APT. In addition, the system is unavailable during the mitigation strategy deployment.

Survivability has been defined by ANSI T1A1.2 committee as the transient performance of a system after an undesirable event [34]. The metrics used to quantify survivability vary according to applications, and depend on a number of factors such as the minimum level of performance necessary for the system to be considered functional and the maximum acceptable security loss of a system. Performance levels are assigned as reward rates. In this paper, we classify the metrics into two broad categories:

- *Instantaneous metrics* are transient metrics that capture the state of the system at time t after the occurrence of an undesired event. An example of an instantaneous metric is the probability that the vulnerable system has been recovered at time t.
- *Cumulative metrics* are integrals of instantaneous metrics, that is, expected accumulated rewards in the interval $(0, t]$.

The metrics considered in this paper include

Metric m_1. Probability that the vulnerable system has been patched at time t;
Metric m_2. Probability that the system is unavailable at time t;
Metric m_3. Mean accumulated time that the system is unavailable in the interval $(0, t]$; and
Metric m_4. Mean accumulated loss of system confidentiality and integrity loss in the interval $(0, t]$.

Note that survivability metrics are transient metrics computed after the announcement of a vulnerability. In the remainder of this paper, time t refers to the time since a vulnerability is found and is measured in days.

3.2   Stochastic Reward Net Model

Figure 2 shows an SRN model for the survivability analysis of a system under an APT attack. Table 1 shows the definition of variables, while Table 2 shows guard definitions. When a software vulnerability is identified, $T_{bugfound}$ fires with a quick rate δ. One token is removed from $p_{bugfound}$ and one token is put in p_{vul_s}, p_{vul}, and $p_{prepare}$ each, representing that system failure, exploitation code implementation and mitigation strategy implementation start concurrently.

When the exploit code is ready, T_{vul} fires. Then, one token is taken from p_{vul} and one token is deposited in $p_{exploit_2}$ and one token is deposited in p_{tvul}. Guard function g_{vul} determines whether or not a token is put in $p_{exploit}$. Only when the system does not fail (represented by a token in p_{vul_s}), t_{vul} fires and a token is put in $p_{exploit}$. That is, the number of tokens in place $p_{exploit_2}$ is used for determining whether the exploit code is ready after the system failure or crash is fixed. Places $p_{exploit}$, p_{infect}, p_{lmov}, and p_{efil} represent the status of the attacker in the system. When $T_{exploit}$ fires, one token is taken from $p_{exploit}$ and one token is put in p_{infect}, representing that exploit code is injected into the system successfully with mean time $1/(\lambda_{exploit} \cdot \rho_1)$. This injection process may fail resulting in the system crash, represented by firing T_{c_1}. For this situation, one token is taken from $p_{exploit}$ and one token is put in p_{crash}. The firing rate is $(1-\rho_1) \cdot \lambda_{exploit}$. In addition, the system may fail due to other reasons, represented by firing T_{f_1} with mean time $1/\lambda_{fail}$. That is, one token is taken from $p_{exploit}$ and one token is put in p_{fail}. Similar explanations are applicable to transitions from p_{infect}, p_{lmov}, and p_{efil}.

A token in place $p_{prepare}$ denotes the condition that the mitigation strategy is under implementation. When $T_{prepare}$ fires, one token is taken from $p_{prepare}$ and one token is put in p_{ready}, representing that the strategy is ready for deployment. When there is a token in p_{ready} and p_{repair} ($p_{vul}, p_{exploit}, p_{infect}, p_{lmov}$, or p_{efil}), the immediate transition t_1 (t_2, t_3, t_4, t_5, or t_6, respectively) fires. Then, a token is taken from p_{ready} and p_{repair} ($p_{vul}, p_{exploit}, p_{infect}, p_{lmov}$, or p_{efil}) and deposited in place p_{deploy}. The place p_{deploy} represents that the system begins the deployment of the mitigation strategy. When T_{deploy} fires, one token is taken from p_{deploy} and one token is put in p_{good}, representing that the system completes the mitigation strategy deployment and enters into state GOOD.

The priority of t_1 over t_7 aims to achieve the following goal: when the mitigation strategy and exploit code are both available, the mitigation strategy must be deployed immediately. Then, t_1 is fired and one token is taken from p_{ready} and p_{repair} each, and one token is put in p_{deploy}. If the strategy is not ready but the exploit code is available, then t_7 is fired. At this time, one token is taken from p_{repair} and $p_{exploit_2}$ each, and one token is put in $p_{exploit}$ and $p_{exploit_2}$. If both are unavailable, then t_7 does not fire and the token is kept in p_{repair}. Later, when mitigation strategy or exploit code is available, transition t_1 or t_7 fires, respectively.

Based on this model, we use the SPNP software package [35] to calculate the four metrics mentioned at the end of Sect. 3.1 as follows:

Table 1. Definition of variables used in monolithic SRN model.

Symbol	Definition	Mean value
$1/\delta$	Mean time that the discovered vulnerability is known to all	30 min
$1/\lambda_{prepare}$	Mean time for implementing a mitigation strategy	20 days
$1/\lambda_{deploy}$	Mean time for installing the mitigation strategy	12 days
$1/\lambda_{vuln}$	Mean time for generating the exploit code	4 days
$1/\lambda_{fail}$	Mean time that the computer system fails	365 days
$1/\lambda_{fix}$	Mean time that the computer system completes the failure or crash fixing	2 days
$1/\lambda_{efil}$	Mean time that the attacker obtains the desired information	2 days
$1/\lambda_{exploit}$	Mean time for injecting the exploit code into the system	7 days
$1/\lambda_{inf}$	Mean time that the exploit code is persistent	1 days
$1/\lambda_{lmov}$	Mean time that the attacker finds sensitive data of interest	7 days
ρ_1	Probability that the exploit code works in the system	0.6
ρ_2	Probability that the exploit code is persistent	0.6
ρ_3	Probability that the attacker finds its target	0.6
ρ_4	Probability that the attacker obtains the desired information	0.6

Table 2. Guard functions for the SRN model.

Guard	Values
g_{vul}	**if** $(\#(p_{vul_s}) == 1)$ **then** 1 **else** 0
g_{f5}	**if** $(\#(p_{vul}) == 1)$ **then** 1 **else** 0

- The value of m_1 at time t is the *expected number of tokens of* p_{good} *at time* t.
- The value of m_2 at time t is the *expected number of tokens of* $(p_{crash} + p_{fail} + p_{deploy})$ *at time* t.
- The value of m_3 in the interval $(0, t]$ is the *expected accumulated reward of* $(p_{crash} + p_{fail} + p_{deploy})$ *by time* t.
- The value of m_4 in the interval $(0, t]$ is the *expected accumulated reward of* p_{exfil} *by time* t.

4   Numerical Results and Discussions

In this section we present the numerical results obtained using SPNP software package [35] to solve the SRN model. In particular, we report the metrics previously described.

$1/\lambda_{prepare}$ is set to 20 days according to [36] and our analysis of vulnerability data set of Google Project Zero security team [37]. Similarly, $1/\lambda_{vuln}$ is set to

4 days according to [38]. Values of the other parameters now are unknown to us and are set based on intuition for sensitivity analysis. Table 1 gives the default values considered of each parameter.

We first investigate the effect of $\lambda_{prepare}$ on both the transient probability of GOOD state and the probability that the system is unavailable (that is, m_1 and m_2 metrics, respectively). Figures 3 and 4 plot these results, respectively. Probabilities ρ_1, ρ_2, ρ_3, and ρ_4 are set to the value denoted by *crash probability*. P04, P08, P12, P16, and P20 represent the results of $1/\lambda_{prepare} = 4$ days, 8 days, 12 days, 16 days, 20 days respectively. We carry out numerical analysis under crash probability of 10 % and 40 %.

From Fig. 3(a) and (b), we observe that crash probability has little effect on the probability of GOOD state for each $\lambda_{prepare}$ in our parameter configurations. The reason is that as long as the mitigation strategy becomes ready, the system can immediately enter into the deployment phase no matter which state of $p_{vuln}, p_{exploit}, p_{infect}, p_{lmov}$, and p_{exfil} is. However, the influence of $\lambda_{prepare}$ value is significant. The larger $\lambda_{prepare}$, the larger increase in the probability of GOOD state at time t. That is, the smaller $1/\lambda_{prepare}$ is, the quicker the system is recovered.

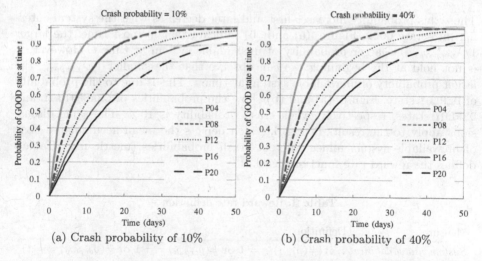

(a) Crash probability of 10%   (b) Crash probability of 40%

Fig. 3. Probability of GOOD state at time t under different crash probabilities (metric m_1).

However, Fig. 4(a) and (b) indicate that both crash probability and $\lambda_{prepare}$ have obvious effects on the probability that the system is unavailable at time t. This probability increases first and then decreases with t. The reason is as follows. At the beginning, only the system failure with very small rate contributes the probability of unavailable system. When the exploitation code is ready, the system crashes frequently. In addition, when the mitigation strategy is ready, the strategy deployment also contributes to the probability of unavailable system.

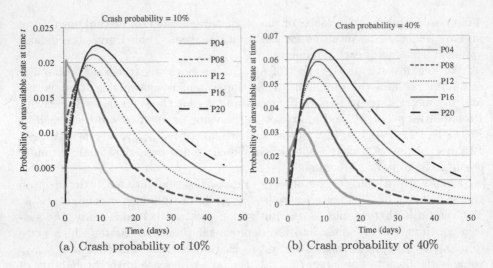

Fig. 4. Probability of unavailable system at time t under different crash probabilities (metric m_2).

Thus, this probability increases first and later decreases with the system entering into state GOOD. Figure 4(a) and (b) indicate that in most time, the larger $1/\lambda_{prepare}$, the larger probability of unavailable system at time t. However, it is not hold at the beginning: we can observe that the smaller $1/\lambda_{prepare}$, the larger probability of unavailable system at time t. This is due to the probability of DEPLOY state. Figure 5(a) and (b) show the probabilities of CRASH+FAIL and DEPLOY states, respectively, when crash probability is 10 %. The same discussions apply to the results in Fig. 6, which depicts the mean accumulated time of unavailable system under different crash probabilities (metric m_3). Table 3 defines the reward rate functions used.

Table 3. Reward rate definition.

Measure	Definition
System unavailability	1: **if** $(\#(p_{fail}) == 1$ **or** $\#(p_{crash}) == 1$ **or** $\#(p_{deploy}) == 1)$
	0: otherwise
Confidentiality loss	1: **if** $(\#(p_{efil}) == 1)$
	0: otherwise

Metric m_4 defined in Sect. 3.1 can be used for computing loss of confidentiality+integrity, integrity, or confidentiality. It depends on the attacker objective. Without loss of generality, we consider the confidentiality in the experiments. The system confidentiality loss per day is defined as 1. Figure 7(a) and (b) plot

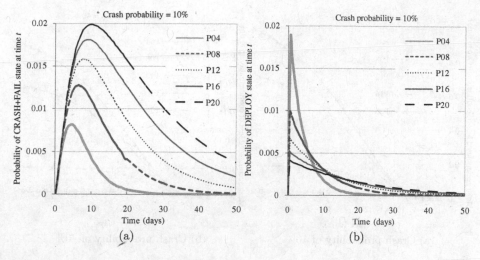

Fig. 5. Probability of (a) `CRASH+FAIL` and (b) `DEPLOY` state at time t under crash probability of 10%.

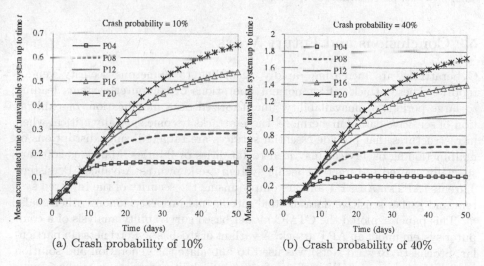

(a) Crash probability of 10%   (b) Crash probability of 40%

Fig. 6. Mean accumulated time that the system is unavailable under different crash probabilities (metric m_3).

the mean accumulated loss of system confidentiality by time t, under different crash probabilities (metric m_4). We can see that the larger $1/\lambda_{prepare}$ and/or the smaller crash probability, the larger mean accumulated time and loss.

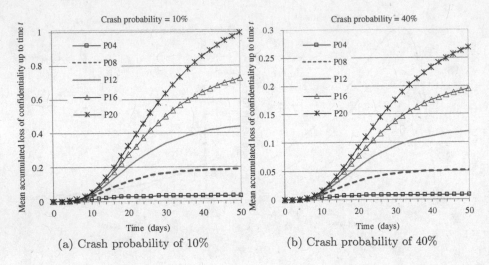

(a) Crash probability of 10%   (b) Crash probability of 40%

Fig. 7. Mean accumulated of system confidentiality and integrity loss by time t under different crash probabilities (metric m_4).

5   Conclusions and Future Work

Cyberattacks are increasing rapidly according to numerous security vendor reports. These attacks affect normal operations of computer systems leading to large periods of unavailability, unauthorized data modification, or exfiltration of sensitive data. Furthermore, these attacks become specially critical when infrastructures that provide essential services are targeted, since disruptions or malfunctioning of services may lead to fatalities or injuries. Malicious software specially crafted to target a specific system are known as Advanced Persistent Threats (APTs). An APT aim at compromising the security of the targeted system and gather sensitive data or steal intellectual property, among other goals.

This paper explored the CTMC model-based survivability analysis of a computer system under an APT attack. A variant of stochastic Petri nets (in particular, Stochastic Reward Nets) was used to automate the generation and solution of the Markov model. We defined four survivability metrics, in terms of system recovery, system availability, data confidentiality loss, and data integrity loss. In addition, numerical results have been presented to study the impact of the underlying parameters on the system survivability. These results may also provide insights on the cost/benefit trade-offs of investment efforts in system recovery strategies including vulnerability mitigation schemes.

There are several future work directions. This paper does not consider the security improvement schemes which could reduce the security loss during the system recovery from a vulnerability. These security schemes include using black-list/whitelist to enforce system access control, using backup software and so on. Extending to our proposed model to capture these schemes and then quantitatively evaluating the abilities of various security protection schemes is our next research. We also plan to extend our survivability-based model to the scenario where multiple vulnerabilities are found and some event times are non-exponentially distributed. Furthermore, the modeling in this paper is an approximation of the real restoration process. We shall extend the current study to approximate the model in a more accurate way to real system behaviors.

Acknowledgments. The research of Ricardo J. Rodríguez was supported by the Spanish MINECO project CyCriSec (TIN2014-58457-R). The research of Xiaolin Chang was supported by NSF 61572066 of China. The research of Xiaodan Li and Kishor S. Trivedi was supported in part by US NSF grant number CNS-1523994, by IBM under a faculty grant, by NATO under Science for Peace project number 984425, and by US Navy under grant N00174-16-C-0036.

References

1. Symantec: Internet Security Threat report (2013). http://www.symantec.com/content/en/us/enterprise/other_resources/bistr_main_report_v18_2012_21291018.en-us.pdf
2. Emm, D., Garnaeva, M., Ivanov, A., Makrushin, D., Unuchek, R.: IT threat evolution in Q2 2015. Technical report, Kaspersky Lab, July 2015
3. McAfee: McAfee labs threats report. Technical report, McAfee Labs, August 2015
4. Department of Homeland Security: National Security Strategy. The White House, May 2010. http://www.whitehouse.gov/sites/default/files/rss_viewer/national_security_strategy.pdf
5. Kozik, R., Choras, M.: Current cyber security threats and challenges in critical infrastructures protection. In: Proceedings of the 2nd International Conference on Informatics and Applications (ICIA), pp. 93–97, September 2013
6. Walters, R.: Cyber Attacks on U.S. Companies in 2014. The Heritage Foundation - National Security and Defense, 1–5, October 2014. Issue Brief No. 4289
7. Moser, A., Kruegel, C., Kirda, E.: Exploring multiple execution paths for malware analysis. In: Proceedings of the IEEE Symposium on Security and Privacy, pp. 231–245 (2007)
8. Bayer, U., Habibi, I., Balzarotti, D., Kirda, E., Kruegel, C.: A view on current malware behaviors. In: Proceedings of the 2nd USENIX Conference on Large-scale Exploits and Emergent Threats: Botnets, Spyware, Worms, and More (LEET), pp. 1–11. USENIX Association, Berkeley (2009)
9. Sood, A., Enbody, R.: Targeted cyberattacks: a superset of advanced persistent threats. IEEE Secur. Priv. **11**(1), 54–61 (2013)
10. Tankard, C.: Advanced persistent threats and how to monitor and deter them. Netw. Secur. **2011**(8), 16–19 (2011)
11. Farwell, J.P., Rohozinski, R.: Stuxnet and the future of cyber war. Survival **53**(1), 23–40 (2011)
12. Rauscher, K.: Writing the rules of cyberwar. IEEE Spectr. **50**(12), 30–32 (2013)
13. Ellison, R.J., Fisher, D.A., Linger, R.C., Lipson, H.F., Longstaff, T.A., Mead, N.R.: Survivability: protecting your critical systems. IEEE Internet Comput. **3**(6), 55–63 (1999)
14. Bolch, G., Greiner, S., de Meer, H., Trivedi, K.S.: Queueing Networks and Markov Chains: Modeling and Performance Evaluation with Computer Science Applications, 2nd edn. Wiley-Interscience, Hoboken (2006)

15. Ramani, S., Trivedi, K.S., Dasarathy, B.: Performance analysis of the CORBA event service using stochastic reward nets. In: Proceedings of the 19th IEEE Symposium on Reliable Distributed Systems (SRDS), pp. 238–247 (2000)
16. Philip, A., Sharma, R.K.: A stochastic reward net approach for reliability analysis of a flexible manufacturing module. Int. J. Syst. Assur. Eng. Manag. 4(3), 293–302 (2013)
17. Bruneo, D.: A stochastic model to investigate data center performance and QoS in IaaS cloud computing systems. IEEE Trans. Parallel Distrib. Syst. 25(3), 560–569 (2014)
18. Entezari-Maleki, R., Trivedi, K.S., Movaghar, A.: Performability evaluation of grid environments using stochastic reward nets. IEEE Trans. dependable Secure Comput. 12(2), 204–216 (2015)
19. Kumar, N., Lee, J.H., Chilamkurti, N., Vinel, A.: Energy-efficient multimedia data dissemination in vehicular clouds: stochastic-reward-nets-based coalition game approach. IEEE Syst. J. 10(2), 847–858 (2016)
20. Kawamura, R., Ohta, H.: Architectures for ATM network survivability and their field deployment. IEEE Commun. Mag. 37(8), 88–94 (1999)
21. Wylie, J.J., Bigrigg, M.W., Strunk, J.D., Ganger, G.R., Kiliccote, H., Khosla, P.K.: Survivable information storage systems. Computer 33(8), 61–68 (2000)
22. Jha, S., Wing, J.M.: Survivability analysis of networked systems. In: Proceedings of the 23rd International Conference on Software Engineering (ICSE), ICSE 2001, pp. 307–317. IEEE Computer Society, Washington, DC (2001)
23. Castet, J.F., Saleh, J.H.: On the concept of survivability, with application to spacecraft and space-based networks. Reliab. Eng. Syst. Saf. 99, 123–138 (2012)
24. Paulauskas, N., Garsva, E., Gulbinovic, L., Stankevicius, A., Poviliauskas, D.: Survivability modelling of Lithuanian government information system. Elektronika Ir Elektrotechnika 120(4), 95–98 (2012)
25. Wang, H., Liu, P.: Modeling and evaluating the survivability of an intrusion tolerant database system. In: Gollmann, D., Meier, J., Sabelfeld, A. (eds.) ESORICS 2006. LNCS, vol. 4189, pp. 207–224. Springer, Heidelberg (2006). doi:10.1007/11863908_14
26. Wang, A.H., Yan, S., Liu, P.: A semi-markov survivability evaluation model for intrusion tolerant database systems. In: Proceedings of the 2010 International Conference on Availability, Reliability, and Security (ARES), pp. 104–111, February 2010
27. Trivedi, K.S., Xia, R.: Quantification of system survivability. Telecommun. Syst. 60(4), 451–470 (2015)
28. Rodríguez, R.J., Merseguer, J., Bernardi, S.: Modelling security of critical infrastructures: a survivability assessment. Comput. J. 58(10), 2313–2327 (2015)
29. Pfleeger, C.P., Pfleeger, S.L.: Security in Computing, 4th edn. Prentice Hall, Upper Saddle River (2006)
30. Murata, T.: Petri nets: properties, analysis and applications. Proc. IEEE 77(4), 541–580 (1989)
31. Ajmone Marsan, M., Balbo, G., Conte, G., Donatelli, S., Franceschinis, G.: Modelling with Generalized Stochastic Petri Nets. Wiley Series in Parallel Computing. Wiley, Hoboken (1995)
32. Muppala, J., Ciardo, G., Trivedi, K.S.: Stochastic reward nets for reliability prediction. Commun. Reliab. Maintainab. Serviceability 1(2), 9–20 (1994)
33. Grottke, M., Trivedi, K.: Fighting bugs: remove, retry, replicate, and rejuvenate. Computer 40(2), 107–109 (2007)

34. ANSI T1A1.2 Working Group on Network Survivability Performance: Enhanced Network Survivability Performance. Technical report 68, American National Standards Institute (2001)
35. Ciardo, G., Muppala, J., Trivedi, K.: SPNP: stochastic Petri net package. In: Proceedings of the 3rd International Workshop on Petri Nets and Performance Models (PNPM), pp. 142–151, December 1989
36. Temizkan, O., Kumar, R., Park, S., Subramaniam, C.: Patch release behaviors of software vendors in response to vulnerabilities: an empirical analysis. J. Manage. Inf. Syst. **28**(4), 305–338 (2012)
37. Google Project Zero: List of vulnerabilities reported by Google security research team. https://bugs.chromium.org/p/project-zero/issues/list?can=1&q=&colspec=ID+Type+Status+Priority+Milestone+Owner+Summary&cells=ids
38. Nzoukou, W., Wang, L., Jajodia, S., Singhal, A.: A unified framework for measuring a network's mean time-to-compromise. In: Proceedings of the 2013 IEEE 32nd International Symposium on Reliable Distributed Systems (SRDS), pp. 215–224, September 2013

Confining Adversary Actions via Measurement

Paul D. Rowe[✉]

The MITRE Corporation, Bedford, USA
prowe@mitre.org

Abstract. Systems designed with measurement and attestation in mind are often layered, with the lower layers measuring the layers above them. Attestations of such systems must report the results of a diverse set of application-specific measurements of various parts of the system. There is a pervasive intuition that measuring the system "bottom-up" (i.e. measuring lower layers before the layers above them) is more robust than other orders of measurement. This is the core idea behind trusted boot processes. In this paper we justify this intuition by characterizing the adversary actions required to escape detection by bottom-up measurement. In support of that goal, we introduce a formal framework with a natural and intuitive graphical representation for reasoning about layered measurement systems.

1 Introduction

Security decisions often rely on trust. Many computing architectures have been designed to help establish the trustworthiness of a system through remote attestation. They gather evidence of the integrity of a target system and report it to a remote party who appraises the evidence as part of a security decision. A simple example is a network gateway that requests evidence that a target system has recently run antivirus software before granting it access to a network. If the virus scan indicates a potential infection, or does not offer recent evidence, the gateway might decide to deny access, or perhaps divert the system to a remediation network. Of course the antivirus software itself is part of the target system, and the gateway may require integrity evidence for the antivirus software for its own security decision. This leads to the design of layered systems in which deeper layers are responsible for generating integrity evidence of the layers above them.

A simple example of a layered system is one that supports "trusted boot" in which a chain of boot-time integrity evidence is generated for a trusted computing base that supports the upper layers of the system. A more complex example might be a virtualized cloud architecture. The virtual machines (VMs) at the top are supported at a lower layer by a hypervisor or virtual machine monitor. Such an architecture may be augmented with additional VMs at an intermediate layer that are responsible for measuring the main VMs to generate integrity evidence. These designs offer exciting possibilities for remote attestation. They allow for specialization and diversity of the components involved, tailoring the capabilities of measurers to their targets of measurement, and composing them in novel ways.

B. Kordy et al. (Eds.): GraMSec 2016, LNCS 9987, pp. 150–166, 2016.
DOI: 10.1007/978-3-319-46263-9_10

However, the resulting layered attestations are typically more complex and challenging to analyze. Given a target system, what set of evidence should an appraiser request? What extra guarantees are provided if it receives integrity evidence of the measurers themselves? Does the order in which the measurements are taken matter?

This paper begins to tame the complexity surrounding attestations of these layered systems. We provide a formal model of layered measurement and attestation systems that abstracts away the underlying details of the measurements and focuses on the causal relationships among component corruption and measurement.

Limitations of Measurement. Our starting point for this paper is the recognition of the fact that measurement cannot *prevent* corruption; at best, measurement only *detects* corruption. In particular, the runtime corruption of a component can occur even if it is launched in a known good state. An appraiser must therefore always be wary of the gap between the time a component is measured and the time at which a trust decision is made. If the gap is large then so is the risk of a time-of-check-to-time-of-use (TOCTOU) attack in which an adversary corrupts a component during the critical time window to undermine the trust decision. A successful measurement strategy will limit the risk of TOCTOU attacks by ensuring the time between a measurement and a security decision is sufficiently small. The appraiser can then conclude that if the measured component is currently corrupted, it must be because the adversary performed a *recent* attack.

Shortening the time between measurement and security decision, however, is effective only if the measurement component can be trusted. By corrupting the measurer, an adversary can lie about the results of measurement making a corrupted target component appear to be in a good state. This affords the adversary a much larger window of opportunity to corrupt the target. The corruption no longer has to take place in the small window between measurement and security decision because the target can already be corrupted at the time of (purported) measurement. However, in a typical layered system design, deeper components such as a measurer have greater protections making it harder for an adversary to corrupt them. This suggests that to escape the burden of performing a *recent* corruption, an adversary should have to pay the price of corrupting a *deep* component.

Formal Model of Measurement and Attestation. With this in mind, our first main contribution is a formal model designed to aid in reasoning about what an adversary must do in order to defeat a measurement and attestation strategy. Rather than forbid the adversary from performing TOCTOU attacks in small windows or from corrupting deep components, we provide results that help to characterize and confine where such undesirable adversary actions must occur if the adversary is to corrupt a component without detection. Thus our model explicitly allows an adversary to corrupt (and repair) arbitrary system components at any time.

The model also features a true concurrency execution semantics which allows us to reason more directly about the causal effects of corruptions on the outcomes of measurement without having to reason about unnecessary interleavings of events. An important side benefit of this semantics is that it admits a natural, graphical representation that helps an analyst quickly understand the causal relationships between events of an execution. This pairs nicely with our analysis method based on characterizing executions consistent with some hypotheses, because it allows an analyst to quickly evaluate these executions without having to specify in advance a particular security goal.

Strategy for Measurement. We demonstrate the utility of this formal model by validating the effectiveness of an important strategy for measurement. An intuition manifest in much of the literature on measurement and attestation is that trust in a system should be based on a bottom-up chain of measurements starting with a hardware root of trust for measurement. This is the core idea behind trusted boot processes, in which one component in the boot sequence measures the next component before launching it. Theorem 1, which we refer to as the "recent or deep" theorem, validates this common intuition and characterizes exactly what an adversary must do to defeat such bottom-up measurement strategies. It roughly says the following:

> If a system has measured deeper components before more shallow ones, then the only way for the adversary to corrupt a component t without detection is either by *recently* corrupting one of t's dependencies, or else by corrupting a component even *deeper* in the system.

Paper Structure. The paper is structured as follows. We motivate our intuitions and informally introduce our model in Sect. 2. In Sect. 3 we formally define the systems of study and their executions. In Sect. 4 we prove some important facts about executions. We also define bottom-up measurement strategies and prove they confine adversary corruptions to be either recent or deep. Section 5 discusses some relevant related work. Finally, we conclude in Sect. 6.

2   Motivating Examples of Measurement

Consider an enterprise that would like to ensure that systems connecting to its network provide a fresh system scan by the most up-to-date virus checker. The network gateway should ask systems to perform a system scan on demand when they attempt to connect. We may suppose the systems all have some component A_1 that is capable of accurately reporting the running version of the virus checker. Because this enterprise values high assurance, the systems also come equipped with another component A_2 capable of measuring the runtime state of the kernel. This is designed to detect any rootkits that might try to undermine the virus checker's system scan. We may assume that A_1 and A_2 are both measured by a hardware root of trust for measurement (rtm) as part of a secure boot process. Thus, the architecture for systems in this enterprise might look

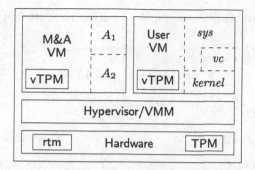

Fig. 1. Example measurement system.

something like Fig. 1 in which the (virtual) trusted platform modules ((v)TPMs) serve to store and report the measurement values to a remote appraiser.

We are thus interested in a system consisting of the following components: $\{sys, vc, ker, A_1, A_2, \mathsf{rtm}\}$, where sys represents the collective parts of the system scanned by the virus checker vc, and ker represents the kernel. Based on the scenario described above, we may be interested in the following set of measurement events

$$\{\mathsf{ms}(\mathsf{rtm}, A_1), \mathsf{ms}(\mathsf{rtm}, A_2), \mathsf{ms}(A_1, vc), \mathsf{ms}(A_2, ker), \mathsf{ms}_{ker}(vc, sys)\}$$

where $\mathsf{ms}_C(o_1, o_2)$ represents the measurement of o_2 by o_1 while C provides the runtime context. These measurement events generate the raw evidence that the network gateway can use to make a determination as to whether or not to admit the system to the network.

If any of the measurements indicate a problem, such as a failed system scan, then the gateway has good reason to believe it should deny the system access to the network. But what if all the evidence it receives looks good? How confident can the gateway be that the version and signature files are indeed up to date? The answer will depend on the order in which the evidence was gathered. The problem of determining the order in which measurements were taken given a set of signed quotes from (v)TPMs is addressed in [12]. In what follows, we assume the appraiser has some way of accurately determining the order in which measurements are taken. To get some intuition for why the order of measurement matters, consider the three different specifications pictured in Fig. 2 (in which time flows from top to bottom) for how to order the measurements. (The bullet after the first three events does not represent a separate event. It is inserted only for visible legibility, to avoid crossing arrows, so that each of the events on the top row occurs before both events on the next row down.)

Specification S_1 ensures that both vc and ker are measured before vc runs its system scan. Specifications S_2 and S_3 each relax one of those ordering requirements. Let's now consider some executions that respect the order of measurements in each of these specifications in which the adversary manages to avoid detection.

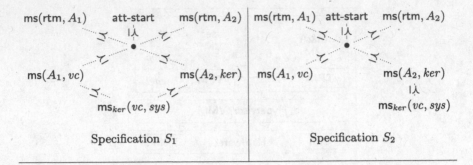

Specification S_1 | Specification S_2

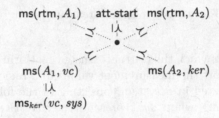

Specification S_3

Fig. 2. Three orders for measurement

Execution E_1 of Fig. 3 is compatible with Specification S_1. The adversary manages to corrupt the system by installing some user-space malware sometime in the past. If we assume the up-to-date virus checker is capable of detecting this malware, then the adversary must corrupt either vc or ker before the virus scan represented by $\mathsf{ms}_{ker}(vc, sys)$. That is, either a corrupted vc will lie about the results of measurement, or else a corrupted ker can undermine the integrity of the system scan, for example, by hiding the directory containing the malware from vc. In the case of E_1, the adversary corrupts vc in order to lie about the results of the system scan, but it does so after $\mathsf{ms}(A_1, vc)$ in order to avoid detection by this measurement event.

In Execution E_2, which is consistent with Specification S_2, the adversary is capable of avoiding detection while corrupting vc much earlier. The system scan $\mathsf{ms}_{ker}(vc, sys)$ is again undermined by the corrupted vc. Since vc will also be measured by A_1, the adversary has to restore vc to an acceptable state before $\mathsf{ms}(A_1, vc)$. Execution E_3 is analogous to E_2, but the adversary corrupts ker instead of vc, allowing it to convince the uncorrupted vc that the system has no malware. Since Specification S_3 allows $\mathsf{ms}(A_2, ker)$ to occur after the system scan, the adversary can leverage the corrupted ker to lie about the scan results, but must restore ker to a good state before it is measured.

Execution E_1 is ostensibly harder to achieve for the adversary than either E_2 or E_3, because the adversary has to work quickly to corrupt vc *during* the attestation. In E_2 and E_3, the adversary can corrupt vc and ker respectively at any time in the past. He still must perform a quick restoration of the corrupted component during the attestation, but there are reasons to believe this may be

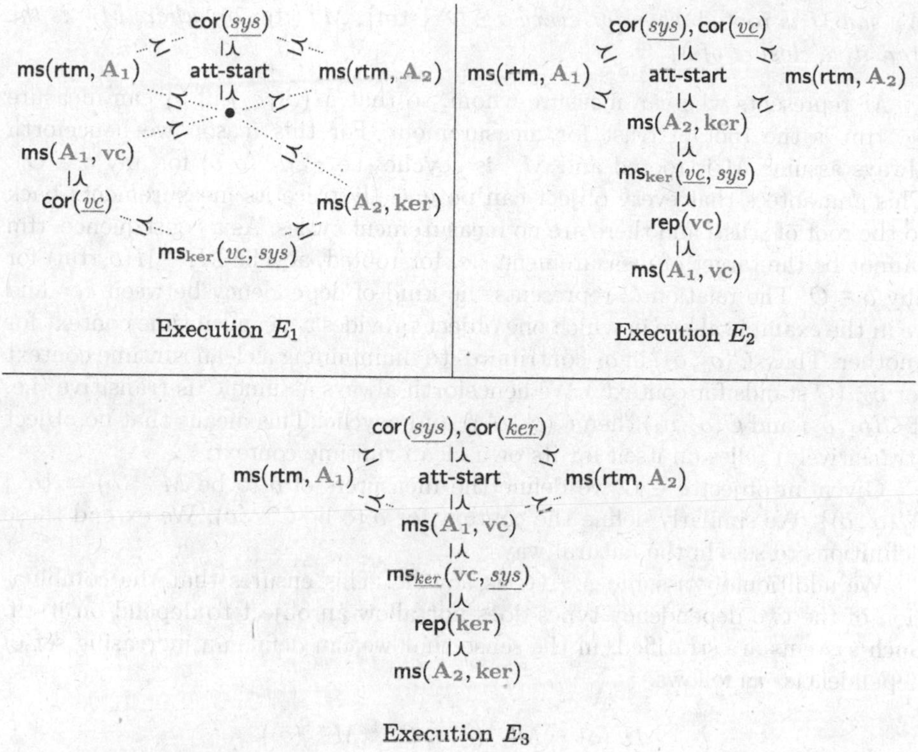

cor(sys)

ms(rtm, A_1)   att-start   ms(rtm, A_2)

ms(A_1, vc)

cor(vc)   ms(A_2, ker)

ms$_{ker}$(vc, sys)

Execution E_1

cor(sys), cor(vc)

ms(rtm, A_1)   att-start   ms(rtm, A_2)

ms(A_2, ker)

ms$_{ker}$(vc, sys)

rep(vc)

ms(A_1, vc)

Execution E_2

cor(sys), cor(ker)

ms(rtm, A_1)   att-start   ms(rtm, A_2)

ms(A_1, vc)

ms$_{ker}$(vc, sys)

rep(ker)

ms(A_2, ker)

Execution E_3

Fig. 3. Three system executions

easier than corrupting the component to begin with. The results of this paper provide a way of characterizing where and when adversary actions must occur in order to avoid detection by measurement. This leads to a result that any execution consistent with S_1 in which the adversary corrupts sys without detection forces the adversary to perform either a recent or a deep corruption.

3   Measurement Systems

In this section we formalize the intuitions we used for the examples in the previous section.

System Architecture. We start by describing the core types of dependencies that make a system layered.

Definition 1 (Measurement Systems). *We define a* measurement system *to be a tuple* $\mathcal{MS} = (O, M, C)$*, where O is a set of objects (e.g. software components) with a distinguished element* rtm*. M and C are binary relations on O. We call*

M the measures *relation, and*
C the context *relation.*

We say M is rooted *when for every $o \in O\backslash\{\text{rtm}\}$, $M^+(\text{rtm}, o)$, where M^+ is the transitive closure of M.*

M represents who can measure whom, so that $M(o_1, o_2)$ iff o_1 can measure o_2. rtm is the root of trust for measurement. For this reason we henceforth always assume M is rooted and M^+ is acyclic (i.e. $\neg M^+(o, o)$ for any $o \in O$). This guarantees that every object can potentially trace its measurements back to the root of trust, and there are no measurement cycles. As a consequence, rtm cannot be the target of measurement, i.e. for rooted, acyclic M, $\neg M(o, \text{rtm})$ for any $o \in O$. The relation C represents the kind of dependency between *ker* and *vc* in the example above in which one object provides a clean runtime context for another. Thus, $C(o_1, o_2)$ iff o_1 contributes to maintaining a clean runtime context for o_2. (C stands for context.) We henceforth always assume C is transitive (i.e. if $C(o_1, o_2)$ and $C(o_2, o_3)$ then $C(o_1, o_3)$) and acyclic. This means that no object (transitively) relies on itself for its own clean runtime context.

Given an object $o \in O$ we define the measurers of o to be $M^{-1}(o) = \{o' \mid M(o', o)\}$. We similarly define the context for o to be $C^{-1}(o)$. We extend these definitions to sets in the natural way.

We additionally assume $M \cup C$ is acyclic. This ensures that the combination of the two dependency types does not allow an object to depend on itself. Such systems are stratified, in the sense that we can define an increasing set of dependencies as follows.

$$D^1(o) = M^{-1}(o) \cup C^{-1}(M^{-1}(o))$$
$$D^{i+1}(o) = D^1(D^i(o))$$

So $D^1(o)$ consists of the measurers of o and their context. As we will see later, $D^1(o)$ represents the set of components that must be uncompromised in order to trust the measurement of o.

We can represent measurement systems pictorially as a graph whose vertices are the objects of $\mathcal{MS}$ and whose edges encode the M and C relations. We use the convention that $M(o_1, o_2)$ is represented by a solid arrow from o_1 to o_2, while $C(o_1, o_2)$ is represented by a dotted arrow from o_1 to o_2. The representation of the system described in Sect. 2 is shown in Fig. 4.

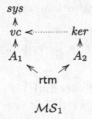

$\mathcal{MS}_1$

Fig. 4. Graphical representation of an example measurement system.

Events, Executions, and Outputs. The components $o \in O$ and the adversary on this system perform actions. In particular, objects can measure each other and the adversary can corrupt and repair components in an attempt to influence the outcome of future measurement actions. Additionally, an appraiser has the ability to inject a random nonce $n \in \mathcal{N}$ into an attestation in order to control the recency of events.

Definition 2 (Events). *Let $\mathcal{MS}$ be a target system. An event for $\mathcal{MS}$ is a node e labeled by one of the following.*

a. *A measurement event is labeled by* $\mathsf{ms}_{C^{-1}(o_2)}(o_2, o_1)$ *such that $M(o_2, o_1)$. We say such an event* measures o_1, *and we call o_1 the* target *of e. When $C^{-1}(o_2)$ is empty we omit the subscript and write* $\mathsf{ms}(o_2, o_1)$.
b. *An* adversary event *is labeled by either* $\mathsf{cor}(o)$ *or* $\mathsf{rep}(o)$ *for $o \in O \backslash \{\mathsf{rtm}\}$.*
c. *The* attestation start *event is labeled by* $\mathsf{att\text{-}start}$.

When an event e is labeled by ℓ we will write $e = \ell$. We will often refer to the label ℓ as an event when no confusion will arise.

An event e touches o *iff o is an argument to the label of e.*

The $\mathsf{att\text{-}start}$ event serves to bound events in time. It represents the choice by the appraiser of a random nonce. Typically the measurements will be cryptographically bound to this nonce before sending them back to the appraiser. In this way, the appraiser will know that anything occurring after this event can reasonably be said to occur "recently". Regarding the measurement events, the rtm is typically responsible for measuring components at boot-time. All other measurements are load-time or runtime measurements of one component in O by another. Adversary events represent the corruption ($\mathsf{cor}(\cdot)$) and repair ($\mathsf{rep}(\cdot)$) of components. Notice that we have excluded rtm from corruption and repair events. This is not because we assume the rtm to be immune from corruption, but rather because all the trust in the system relies on the rtm: Since it roots all measurements, if it is corrupted, none of the measurements of other components can be trusted.

As we saw in the motivational examples, an execution can be described as a partially ordered set (poset) of these events. We choose a partially ordered set rather than a totally ordered set because the latter unnecessarily obscures the difference between *causal* orderings and *coincidental* orderings. However, due to the causal relationships between components, we must slightly restrict our partially ordered sets in order to make sense of the effect that corruption and repair events have on measurement events. To that end, we next introduce a sensible restriction to these partial orders.

A poset is a pair $(E, \prec)$, where E is any set and $\prec$ is a transitive, acyclic relation on E. When no confusion arises, we often refer to $(E, \prec)$ by its underlying set E and use $\prec_E$ for its order relation. Given a poset $(E, \prec)$, let $e{\downarrow} = \{e' \mid e' \prec e\}$, and $e{\uparrow} = \{e' \mid e \prec e'\}$. Given a set of events E, we denote the set of adversary events of E by $adv(E)$ and the set of measurement events by $meas(E)$.

Let $(E, \prec)$ be a partially ordered set of events for $\mathcal{MS} = (O, M, C)$ and let $(E_o, \prec_o)$ be the substructure consisting of all and only events that touch o. We say $(E, \prec)$ is *adversary-ordered* iff for every $o \in O$, $(E_o, \prec_o)$ has the property that if e and e' are incomparable events, then neither e nor e' are adversary events.

Lemma 1. *Let* $(E, \prec)$ *be a finite, adversary-ordered poset for* MS, *and let* $(E_o, \prec_o)$ *be its restriction to some* $o \in O$. *Then for any non-adversarial event* $e \in E_o$, *the set* $adv(e{\downarrow}) \cap E_o$ *is either empty or has a unique maximal element.*

Proof. Since $(E, \prec)$ is adversary-ordered, $adv(E_o)$ is partitioned by $adv(e{\downarrow})$ and $adv(e{\uparrow})$. Suppose $e{\downarrow}$ is not empty. Then since E_o is finite, it has at least one maximal element. Suppose e' and e'' are distinct maximal elements. Thus they must be $\prec_o$-incomparable. However, since $(E, \prec)$ is adversary-ordered, either $e' \prec_o e''$ or $e'' \prec_o e'$, yielding a contradiction.   □

Definition 3 (Corruption State). *Let* $(E, \prec)$ *be a finite, adversary-ordered poset for* MS. *For each event* $e \in E$ *and each object* o *the* corruption state *of* o *at* e, *written* $cs(e, o)$, *is an element of* $\{\bot, r, c\}$ *and is defined as follows.* $cs(e, o) = \bot$ *iff* $e \notin E_o$. *Otherwise, we define* $cs(e, o)$ *inductively:*

$$
cs(e, o) = \begin{cases}
c & : e = cor(o) \\
r & : e = rep(o) \\
r & : e \in meas(E) \wedge adv(e{\downarrow}) \cap E_o = \emptyset \\
cs(e', o) & : e \in meas(E) \wedge e' \text{ maximal in } adv(e{\downarrow}) \cap E_o
\end{cases}
$$

When $cs(e, o)$ *takes the value* c *we say* o *is* corrupt *at* e; *when it takes the value* r *we say* o *is* uncorrupt *or* regular *at* e; *and when it takes the value* $\bot$ *we say the corruption state is* undefined.

We now define what it means to be an execution of a measurement system.

Definition 4 (Executions). *An* exectuion *of a measurement system* MS *is any finite, adversary-ordered poset* E *for* MS.

Since executions are finite and adversary-ordered, for every o, we can always determine the corruption state of o at every event e touching o. We can therefore use these corruption states to determine the outputs of measurements. Abstractly, we assume that for each target of measurement o_t, every measurer of o_t outputs values within some set $MV(o)$.

The question of measurement accuracy is complicated because there are two primary sources of inaccuracy. First, the measurer may not produce values that strongly correlate to the corruption state of the target. For example, an asset inventory tool may only output the version numbers of software installed, and this cannot detect undiscovered (and thus unpatched) vulnerabilities. Second, the appraiser is ultimately the one to *interpret* the output. That is, the appraiser partitions $MV(o)$ into $G(o)$ and $B(o)$. The first set $G(o)$ represents measurement values the appraiser believes represent an uncompromised component, while $B(o)$ are those values the appraiser believes represent a corrupted component. Thus the composition of measurement output with appraiser interpretation forms a classifier for the corruption state of the target whose false positive and negative rates depend on both the measurer and the appraiser.

In this work, to simplify the analysis, we assume there are no false positives or negatives as long as the measurer and its context are uncorrupted. However, we assume a corrupted measurer (or its context) can always convince the appraiser that the target of measurement is uncorrupted.

Assumption 1 (Measurement Accuracy). Let $\mathcal{G}(o)$ and $\mathcal{B}(o)$ be a partition for $\mathcal{MV}(o)$. Let $e = \mathsf{ms}(o_2, o_1)$. The *output* of e, written $out(e)$, is defined as follows.

$$out(e) = \begin{cases} v \in \mathcal{B}(o_1) & cs(e, o_1) = \mathsf{c} \text{ and } \forall o \in \{o_2\} \cup C^{-1}(o_2) \,.\, cs(e, o) = \mathsf{r} \\ v \in \mathcal{G}(o_1) & \text{otherwise} \end{cases}$$

If $out(e) \in \mathcal{B}(o_1)$ we say e *detects a corruption*. If $out(e) \in \mathcal{G}(o_1)$ but $cs(e, o_1) = \mathsf{c}$, we say the adversary *avoids detection at* e.

Given an execution E, Assumption 1 says we can always determine the appraiser's classification. However, it can also be used to infer the corruption states of some components given the corruption states of others and the classification. That is, suppose we know the adversary avoids detection at $e = \mathsf{ms}_{C^{-1}(o)}(o, o_t)$. Then we can conclude that at least one member of $\{o\} \cup C^{-1}(o)$ is corrupt at e. This is an important inference for our main result.

One can imagine weakening Assumption 1 to account for imperfect classification. For example, it would be interesting to perform a probabilistic analysis accounting for false positive and negative rates. However, we leave such investigations for future work.

Although executions always allow us to infer the corruption state of components at events and the outputs of measurements, this only holds if we have accounted for all the adversary actions. The main goal of our framework is to allow an appraiser to infer what adversary events must have occurred and when, assuming some basic facts about an execution. To that end we introduce specifications, which formalize the partial knowledge an appraiser has about the execution of the system.

Definition 5 (Specifications). *A* specification *for measurement system* $\mathcal{MS}$, *is a finite adversary-ordered poset* S *with some (possibly empty) set of assumptions about measurement events regarding*

1. *the corruption states of some of their arguments, or*
2. *the output classification ($\mathcal{G}(o)$ or $\mathcal{B}(o)$).*

A specification S admits *execution* E *iff there is an injective, label-preserving map of partial orders* $\alpha : S \to E$ *preserving assumptions on corruption states and output classifications. The set of all executions admitted by* S *is denoted* $\mathcal{E}(S)$.

We can annotate our diagrams in order to convey the assumptions about measurement events. In particular, we underline the corrupted components and display them in red, and we use bold typeface for the uncorrupted components and display them in green. We can also annotate measurements with a ✓ (resp. X) to indicate the output is in $\mathcal{G}(o)$ (resp. $\mathcal{B}(o)$). This allows for a quite compact graphical representation of the relevant information as seen, for example, in Figs. 3 and 5. We omit these visual annotations from executions when the diagram is too cluttered because they can be inferred.

4   Confining Adversary Behavior

In this section we explore what an appraiser can infer about $\mathcal{E}(S)$, given a specification S. In particular, we are interested in characterizing the ways in which an adversary can corrupt a component without the execution giving any indication of corruption. We thus start with some simple general results about executions, before presenting our main theorem.

Lemma 2. *Let E be an execution of $\mathcal{MS}$, and let e be an event of E touching o. If $cs(e, o) = $ c, then there is a most recent corruption event $e' \preceq e$.*

Proof. This follows immediately from Lemma 1 and Definition 3.   □

This lemma is useful for inferring the existence of a corruption event *before* some given event e. However, since we are also interested in the recency of corruption, we would like to infer the existence of a corruption event *after* a given event. The following lemma allows us to do just that.

Lemma 3. *Let E be an execution of $\mathcal{MS}$, and let $e_1 \prec e_2$ be measurement events touching o such that $cs(e_1, o) = $ r *(resp.* c*)* and $cs(e_2, o) = $ c *(resp.* r*)*. Then there exists a corruption (resp. repair) event e' such that $e_1 \prec e' \prec e_2$.*

Proof. Let $A_1 = adv(e_1\downarrow) \cap E_o$ and $A_2 = adv(e_2\downarrow) \cap E_o$. Since $e_1 \prec e_2$, $A_1 \subseteq A_2$. By Lemma 1, A_2 is either empty or has a unique maximum. However, it can't be empty because then A_1 would also be empty and Definition 3 would imply that $cs(e_1, o) = cs(e_2, o) = $ r, contrary to the hypothesis. So let e' be the unique maximum of A_2. Since $(E, \prec)$ is adversary ordered, either $e' \prec e_1$ or $e_1 \prec e'$. In the first case, e' would also be a maximum of A_1 since $A_1 \subseteq A_2$. But this would imply $cs(e_1, o) = cs(e', o) = cs(e_2, o)$ violating our assumption. Thus $e_1 \prec e' \prec e_2$, and since the corruption state of o is different at e_1 and e_2, e' must change the corruption state.   □

We now turn to a formalization of the rule of thumb at the end of Sect. 2. In particular, we characterize what a bottom-up measurement strategy guarantees. That is, if whenever o_1 depends on o_2 we measure o_2 before measuring o_1, then we seek to understand the constraints this puts on the adversary actions in order to avoid detection. For this discussion we fix a target system $\mathcal{MS}$. Recall that $D^1(o)$ represents the measurers of o and their runtime context.

Definition 6. *A measurement event* $e = \mathsf{ms}(o_2, o_1)$ *in execution* E *is* well-supported *iff either*

i. $o_2 = \mathsf{rtm}$, *or*
ii. *for every* $o \in D^1(o_1)$, *there is a measurement event* $e' \prec_E e$ *such that* o *is the target of* e'.

When e *is well-supported, we call the set of* e' *from Condition* ii *above the* support *of* e. *An execution* E *measures* bottom-up *iff each measurement event* $e \in E$ *is well-supported.*

Theorem 1 (Recent or Deep). *Let* E *be an execution with well-supported measurement event* $e = \mathsf{ms}(o_1, o_t)$ *where* $o_1 \neq \mathsf{rtm}$. *Suppose that* E *detects no corruptions. If the adversary avoids detection at* e, *then either*

1. *there exist* $o \in D^1(o_t)$ *and* $o' \in M^{-1}(o)$ *such that* $\mathsf{ms}(o', o) \prec_E \mathsf{cor}(o) \prec_E e$
2. *there exists* $o \in D^2(o_t)$ *such that* $\mathsf{cor}(o) \prec_E e$.

Proof. Since the adversary avoids detection at e, o_t is corrupt at e, and by Assumption 1, there is some $o \in \{o_1\} \cup C^{-1}(o_1) \subseteq D^1(o_t)$ that is also corrupt at e. Also, since e is well-supported, and $o_1 \neq \mathsf{rtm}$, we know there exists $e' = \mathsf{ms}(o', o)$ with $e' \prec_E e$. We now take cases on $cs(e', o)$.

If $cs(e', o) = \mathsf{r}$ then we apply Lemma 3 to conclude there must be a corruption $\mathsf{cor}(o)$ between e' and e satisfying Clause 1.

If $cs(e', o) = \mathsf{c}$, then since E detects no corruptions, then by Assumption 1, there must be some $o^* \in \{o'\} \cup C^{-1}(o') \subseteq D^2(o_t)$ such that $cs(e', o^*) = \mathsf{c}$. We then apply Lemma 2 to infer there must be a previous corruption $\mathsf{cor}(o^*) \prec_E e' \prec_E e$ satisfying Clause 2. □

This theorem says, roughly, that if measurements indicate things are good when they are not, then there must either be a recent corruption or a deep corruption. This tag line of "recent or deep" is particularly apt if (1) the system dependencies also reflect the relative difficulty for an adversary to corrupt them, and (2) the higher level measurements occur after the att-start event. By ordering the measurements so that more robust ones are measured first, it means that for an adversary to avoid detection for an easy compromise, he must have compromised a measurer recently (i.e. since it itself was measured and typically after the att-start event), or else, he must have previously (though not necessarily recently) compromised a more robust component. In this way, the measurement of a component can raise the bar for the adversary. If, for example, a measurer sits in a privileged location outside of some VM containing a target, it means that the adversary would also have to break out of the target VM and compromise the measurer to avoid detection. The skills and time necessary to perform such an attack are much greater than simply compromising the end target.

Let's illustrate this result in the context of the example of Sect. 2. Consider the specification S' in Fig. 5, which is like specification S_1 from Fig. 2, except that it is annotated with more assumptions in order to satisfy all the hypotheses of Theorem 1. Since these facts are (by definition) preserved by homomorphisms

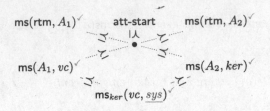

Fig. 5. Specification S'.

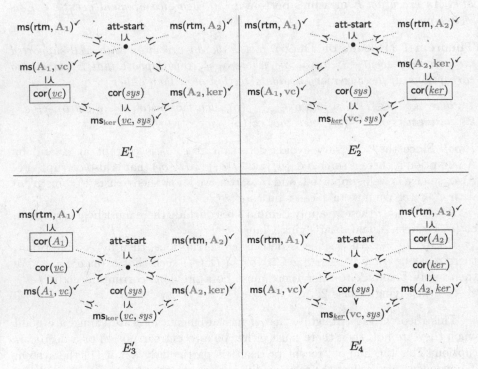

Fig. 6. Executions in $\mathcal{E}(S')$ that do not detect corruption of sys.

$\alpha : S \to \mathcal{E}(S)$, all executions in $\mathcal{E}(S)$ must satisfy them too. Execution E_1 illustrates an example of the first clause of the conclusion being satisfied. There is a "recent" corruption of vc in the sense that vc is corrupted after it is measured. Since the measurement of vc occurs after the start of the attestation, this is truly recent, in that the adversary has very little time to work. The appraiser can control this by ensuring that attestations time out after some fixed amount of time.

Theorem 1 also indicates other possible executions in which the adversary can undetectably corrupt sys. There could be a recent corruption of vc, or else there could be some previous corruption of either A_1 or A_2. All the various options are shown in Fig. 6 in which the corruption events guaranteed by the theorem are boxed. Our theorem allows us to know that these executions essentially characterize all the cases in which a corrupted *sys* goes undetected.

Automation. The analysis above was performed by hand. It would be possible to automate the reasoning steps codified by Assumption 1 and Lemmas 2 and 3. An automated algorithm would have to implement the process of building an execution consistent with (a) the reasoning principles laid out above, and (b) the initial assumptions given by a specification. This is an instance of the more general problem of model finding. That is, given a logical theory and a set of assumptions about a structure, model finding techniques can produce a set of models consistent with the theory and the assumptions.

General purpose tools have been developed that can automate the model finding process [8,13]. We have not yet attempted to use these tools for the analysis of measurement systems. As such, it is unclear if the general algorithms they use will yield efficient analyses, or if they will suffer from combinatorial state space explosions. We have also not investigated the computational complexity of finding a minimal set of executions consistent with a given specification. This would be an interesting question for future work. It is worth noting, however, that Theorem 1 obviates the need to perform such case-by-case analyses when the specification in question is already bottom-up. The value of automated algorithms is greatest when an analyst is unable to apply Theorem 1, for example if the measurement system has cycles in $M \cup C$. Such cyclic dependencies are surprisingly common in production systems because the necessary isolation provided by hardware virtualization is still relatively rare. Thus we believe an automated tool implementing the reasoning principles presented in this paper would be a valuable asset for the analysis of layered attestations.

5 Related Work

There has been much research into measurement and attestation. While a complete survey is infeasible for this paper, we mention the most relevant highlights in order to describe how the present work fits into the larger context of research in this area.

Much of the early work on measurement and attestation was focused on techniques for measuring low-level components that make up a trusted computing base (TCB). These ideas have matured into implementations such as Trusted Boot [11]. Recognizing that many security failures cannot be traced back to the TCB, Sailer et al. [14] proposed an integrity measurement architecture (IMA) in which each application is measured (by hashing its code) before it is launched. More recently, there has been work trying to identify and measure dynamic properties of system components in order to create a more comprehensive picture of the runtime state of a system [5,9,10,15]. All these efforts try to establish what evidence is useful for inferring system state relevant to security decisions. The present work takes for granted that such special purpose measurements can be taken and that they will accurately reflect the system state. Rather, our focus is on developing principles for how to combine a variety of these measurers in a layered attestation. We envision a system designer choosing the measurement capabilities that best suit her needs and using our work to ensure an appraiser can trust the integrity of the result.

In [4], Datta et al. introduce a formalism that accounts for actions local to the target machine as well as network events such as sending and receiving messages. Although they give a very careful treatment of the effect of a corrupted component on an attestation, their work differs in two key ways. First, the formalism represents many low-level details making their proof rather complex, sometimes obscuring the underlying principles. Second, their framework only accounts for static corruptions, while ours is specifically designed around the possibility of dynamic corruption and repair of system components.

Cabuk et al. [1] have proposed an architecture designed to support layered platforms with hierarchical dependencies. It introduces trusted software into the TCB as a software-based root of trust for measurement (SRTM). Although they explain how measurements by the SRTM integrate with the chain of measurements stored in a TPM, they do not study the effect corruptions of various components have on the outcome of attestations. In [2], Coker et al. identify five guiding principles for designing an architecture to support remote attestation. They also describe the design of a (layered) virtualized system based on these principles, although there does not appear to be a publicly available implementation at the time of writing. Of particular interest is a section that describes a component responsible for managing attestations. The emphasis is on the mechanics of selecting measurement agents by matching the evidence they can generate to the evidence requested by an appraiser. There is no discussion or advice regarding the relative order of measurements or the creation of an evidence bundle to reflect the order. More recently, modular attestation frameworks instantiating [2]'s principles have been implemented [3, 6, 7]. These are integrated frameworks that offer plug-and-play capabilities for measurement and attestation for specific usage scenarios. It is precisely these types of systems (in implementation or design) to which our analysis techniques would be most useful. We have not been able to find a discussion of the potential pitfalls of misconfiguring these complex systems. Our work should be able to help guide the configuration of such systems and analyze particular attestation scenarios for each architecture.

Finally, we mention a companion paper to the present work [12]. While the present work helps us characterize how a given measurement order can confine the actions of an adversary, it does not address the question of how a remote appraiser learns the order and results of measurements. This is typically done by storing the evidence in a trusted platform module (TPM) and quoting the results. In [12] it is shown that some methods of using a TPM allow an adversary to bypass Theorem 1 by convincing the appraiser that measurements were taken bottom-up when in fact they were not. The main result is a proposed method for storing and reporting evidence using TPMs ensuring that if an adversary successfully avoids the hypothesis of Theorem 1, then he must nonetheless submit himself to its conclusions.

6   Conclusion

In this paper we have developed a formalism for reasoning about measurement in layered systems. Within this framework we have demonstrated some reusable principles for inferring properties of adversary actions in executions, and we have applied those principles to justify the intuition (pervasive in the literature on measurement and attestation) that it is important to measure a layered system from the bottom up (Theorem 1). Our model admits natural graphical representations of measurement systems, specifications and executions. We believe this graphical representation makes the formalism more intuitive to use, as it allows an analyst to apply her intuitions more immediately to the diagrams.

We believe the model is also relatively extensible in that further types of components and events could be added without disrupting the current results. Indeed, in our companion paper [12], we add events for interacting with a Trusted Platform Module (TPM) in order to perform a more complete analysis of how not just the outcomes of measurements but also their order can be conveyed to a remote appraiser. This is a crucial part of an end-to-end analysis as the appraiser cannot directly observe the order of measurements.

In future work, we would like to consider relaxing Assumption 1 to allow for some probabilistic errors in the classification of measurement targets. Disentangling how much of those errors is due to inaccurate measurement and how much to inaccurate interpretation of the measurement could yield more fine grained results that allow a more nuanced risk decision on the part of the appraiser. We also believe the model could benefit from tool support. The basic problem of discovering adversary actions given assumptions on an execution can be viewed as an instance of model finding. As such, tools such as [8,13] that have been developed for that purpose could be applicable here.

Acknowledgments. I would like to thank Pete Loscocco for suggesting and guiding the direction of this research. Many thanks also to Perry Alexander and Joshua Guttman. Their valuable feedback on during the formation of these ideas was invaluable. Thanks also to Sarah Helble and Aaron Pendergrass for lively discussions about implementations of measurement and attestation systems. Finally, I would like to thank the anonymous reviewers as well as the GraMSec participants for their insightful comments and suggestions for improving the paper.

References

1. Cabuk, S., Chen, L., Plaquin, D., Ryan, M.: Trusted integrity measurement and reporting for virtualized platforms. In: Chen, L., Yung, M. (eds.) INTRUST 2009. LNCS, vol. 6163, pp. 180–196. Springer, Heidelberg (2010)
2. Coker, G., Guttman, J.D., Loscocco, P., Herzog, A.L., Millen, J.K., O'Hanlon, B., Ramsdell, J.D., Segall, A., Sheehy, J., Sniffen, B.T.: Principles of remote attestation. Int. J. Inf. Secur. 10(2), 63–81 (2011)
3. Intel Corporation: Open attestation. Accessed 16 Dec 2015

4. Datta, A., Franklin, J., Garg, D., Kaynar, D.K.: A logic of secure systems and its application to trusted computing. In: 30th IEEE Symposium on Security and Privacy (S&P 2009), Oakland, California, USA, 17–20 May 2009, pp. 221–236 (2009)
5. Davi, L., Sadeghi, A.-R., Winandy, M.: Dynamic integrity measurement, attestation: towards defense against return-oriented programming attacks. In: Proceedings of the 4th ACM Workshop on Scalable Trusted Computing, STC 2009, Chicago, Illinois, USA, 13 November 2009, pp. 49–54 (2009)
6. Fisher, C., Bukovick, D., Bourquin, R., Dobry, R.: SAMSON - Secure Authentication Modules. Accessed 16 Dec 2015
7. Trusted Computing Group. TCG Trusted Network Connect Architecture for Interoperability version 1.5 (2012)
8. Jackson, D.: Software Abstractions: Logic Language and Analysis, 2nd edn. MIT Press, Cambridge (2012)
9. Kil, C., Sezer, E.C., Azab, A.M., Ning, P., Zhang, X.: Remote attestation to dynamic system properties: towards providing complete system integrity evidence. In: Proceedings of the IEEE/IFIP International Conference on Dependable Systems and Networks, DSN 2009, Estoril, Lisbon, Portugal, 29 June–2 July 2009, pp. 115–124 (2009)
10. Loscocco, P., Wilson, P.W., Pendergrass, J.A., McDonell, C.D.: Linux kernel integrity measurement using contextual inspection. In: Proceedings of the 2nd ACM Workshop on Scalable Trusted Computing, STC 2007, Alexandria, VA, USA, 2 November 2007, pp. 21–29 (2007)
11. Maliszewski, R., Sun, N., Wang, S., Wei, J., Qiaowei, R.: Trusted boot (tboot). Accessed 16 Dec 2015
12. Rowe, P.D.: Bundling evidence for layered attestation. In: Franz, M., Papadimitratos, P. (eds.) TRUST 2016. LNCS, vol. 9824, pp. 119–139. Springer, Heidelberg (2016). doi:10.1007/978-3-319-45572-3_7
13. Saghafi, S., Dougherty, D.J.: Razor: provenance and exploration in model-finding. In: 4th Workshop on Practical Aspects of Automated Reasoning (PAAR) (2014)
14. Sailer, R., Zhang, X., Jaeger, T., van Doorn, L.: Design and implementation of a TCG-based integrity measurement architecture. In: Proceedings of the 13th USENIX Security Symposium, San Diego, CA, USA, 9–13 August 2004, pp. 223–238 (2004)
15. Wei, J., Calton, P., Rozas, C.V., Rajan, A., Zhu, F.: Modeling the runtime integrity of cloud servers: a scoped invariant perspective. In: Cloud Computing, Second International Conference, CloudCom 2010, Indianapolis, Indiana, USA, Proceedings, 30 November–3 December 2010, pp. 651–658 (2010)

Author Index

Printed in the United States
By Bookmasters